Aliaksandr Stepanovič

Development of Porous Photoelectrodes for Solar Water Splitting

Aliaksandr Stepanovič

Development of Porous Photoelectrodes for Solar Water Splitting

Combinatorial Approach

Südwestdeutscher Verlag für Hochschulschriften

Impressum / Imprint
Bibliografische Information der Deutschen Nationalbibliothek: Die Deutsche Nationalbibliothek verzeichnet diese Publikation in der Deutschen Nationalbibliografie; detaillierte bibliografische Daten sind im Internet über http://dnb.d-nb.de abrufbar.
Alle in diesem Buch genannten Marken und Produktnamen unterliegen warenzeichen-, marken- oder patentrechtlichem Schutz bzw. sind Warenzeichen oder eingetragene Warenzeichen der jeweiligen Inhaber. Die Wiedergabe von Marken, Produktnamen, Gebrauchsnamen, Handelsnamen, Warenbezeichnungen u.s.w. in diesem Werk berechtigt auch ohne besondere Kennzeichnung nicht zu der Annahme, dass solche Namen im Sinne der Warenzeichen- und Markenschutzgesetzgebung als frei zu betrachten wären und daher von jedermann benutzt werden dürften.

Bibliographic information published by the Deutsche Nationalbibliothek: The Deutsche Nationalbibliothek lists this publication in the Deutsche Nationalbibliografie; detailed bibliographic data are available in the Internet at http://dnb.d-nb.de.
Any brand names and product names mentioned in this book are subject to trademark, brand or patent protection and are trademarks or registered trademarks of their respective holders. The use of brand names, product names, common names, trade names, product descriptions etc. even without a particular marking in this works is in no way to be construed to mean that such names may be regarded as unrestricted in respect of trademark and brand protection legislation and could thus be used by anyone.

Coverbild / Cover image: www.ingimage.com

Verlag / Publisher:
Südwestdeutscher Verlag für Hochschulschriften
ist ein Imprint der / is a trademark of
OmniScriptum GmbH & Co. KG
Heinrich-Böcking-Str. 6-8, 66121 Saarbrücken, Deutschland / Germany
Email: info@svh-verlag.de

Herstellung: siehe letzte Seite /
Printed at: see last page
ISBN: 978-3-8381-3963-0

Zugl. / Approved by: Bochum, RUB, Diss., 2013

Copyright © 2014 OmniScriptum GmbH & Co. KG
Alle Rechte vorbehalten. / All rights reserved. Saarbrücken 2014

Dedicated to all scientists, engineers and entrepreneurs who support environmentally friendly and sustainable technologies.

Acknowledgments

The present work was conducted at the Institute for Materials, Faculty of Mechanical Engineering at the Ruhr-Universität Bochum in the timeframe from December 2010 to November 2013.

I would like to acknowledge the International Max-Planck Research School for Surface and Interface Engineering in Advanced Materials (IMPRS-SurMat) for the financial support and wonderful curriculum.

I would like to express my profound gratitude to my supervisor Prof. Dr.-Ing. Alfred Ludwig, for giving me this opportunity to work in his group on a very interesting and important topic, for his scientific support, as well as for a nice atmosphere during our meetings and discussions. In addition, I express many thanks to all members of the group, who assisted me during my research, shared their experience and knowledge with me, – it was extremely important especially when learning how to work with new equipment. Special thanks to Alan Savan and Robert Meyer for technical support, to Dr. Chinmay Khare and Dr. Pio John Buenconsejo for fruitful collaboration, kind consultations and a nice atmosphere in the office. Dr. Chinmay Khare is especially acknowledged for experimental support with GLAD depositions. Dr. Pio John Buenconsejo is especially acknowledged for the help with TEM measurements. I express my thanks to Alan Savan for checking my dissertation as a native English speaker. Also I would like to thank my PhD colleague Sara Borhani Haghighi for being so helpful and full of positive energy.

I am also thankful to my co-supervisor Prof. Dr. Wolfgang Schuhmann and his group members, especially Kirill Sliozberg, for fruitful discussions and experimental support with photoelectrochemical measurements.

I want to thank Dr. Christoph Somsen for such important help with official paperwork and the hospitable reception on the day of my arrival to Bochum. Also I am thankful to Dr. Rebekka Loschen, Elke Gattermann and Dr. Andreas Erbe for their support in all matters concerning SurMat.

Finally, I would like to thank my family and friends for their moral support.

Abstract

In the present thesis, porous thin film electrodes for photoelectochemical (PEC) solar water splitting applications were developed by the dealloying approach. Combinatorial and high-throughput methods were used for fabrication and investigation of the thin film materials libraries. First, porous metallic precursor films (W, Ti) were synthesized by dealloying sputtered W-Fe, W-Al, W-Mg, Ti-Mg and W-Ti-Mg thin film materials libraries with continuous composition and thickness gradients by etching them in aqueous HNO_3 solutions with different concentrations under open circuit conditions. During the dealloying process, the electrochemically more active elements (Fe, Al and Mg) are selectively dissolved and the remaining W or Ti form porous films with different morphologies. In the following process step these dealloyed films were transformed into porous WO_3 and TiO_2 films by thermal oxidation at 500°C in air. These oxidized films consist of a bilayer structure with a porous upper layer and a dense, not-dealloyed layer underneath. For high-throughput determination of elemental composition of the thin-film materials libraries energy dispersive X-ray (EDX) analysis was used. High-throughput X-ray diffraction (XRD) analysis was used to identify crystalline phases in as-deposited, dealloyed and oxidized films. Film thickness was determined by measuring the height of the specially fabricated steps between a film and a substrate with a stylus profilometer. Investigation of the film morphology was performed by scanning electron microscopy (SEM), using secondary electron and in-lens detectors. Transmission electron microscopy (TEM) was used for morphology and structure characterization with high spatial resolution. PEC properties of the fabricated porous electrodes were investigated in dependence on film thickness, precursor composition and dealloying conditions by means of a scanning droplet cell with an integrated optical fiber. High-throughput characterization of the PEC properties of these porous oxide electrode materials shows a strong dependence of photocurrent densities on the thickness of the dense layer and its composition. Photocurrent density increases with decreasing thickness of the dense layer and Fe, Al or Mg content in it. Nevertheless the dense layer provides mechanical stability of the film and may not be completely eliminated. A new approach was found to provide mechanical stability of the dealloyed film and minimize the negative influence of the dense layer by substituting it with columnar structures, grown by glancing angle sputtering. Photocurrent densities measured at the porous WO_3 electrodes prepared by dealloying are 30-40 times higher than at dense WO_3 thin film electrodes.

Contents

1 **Introduction**......1
  1.1 Solar-hydrogen technologies for sustainable energy supply......1
  1.2 Materials for solar water splitting......3
  1.3 Aims of the thesis......6

2 **Fundamentals**......8
  2.1 Chemical and electrochemical corrosion......8
  2.2 Selective corrosion of alloys......14
    2.2.1 Kinetics of alloy corrosion......15
    2.2.2 Thermodynamics of binary alloy corrosion......16
  2.3 Combinatorial materials science......19

3 **Experimental methods**......22
  3.1 Preparation of thin film materials libraries......22
    3.1.1 Magnetron sputter deposition......22
    3.1.2 Dealloying......26
  3.2 Characterization......27
    3.2.1 Compositional analysis – energy dispersive X-ray analysis (EDX)......27
    3.2.2 Structural analysis – X-ray diffraction and X-ray scattering techniques......28
    3.2.3 Film thickness measurements......29
    3.2.4 Scanning electron microscopy (SEM)......29
    3.2.5 Transmission electron microscope (TEM)......30
    3.2.6 Photoelectrochemical (PEC) measurements......31

4 **Results and discussion**......38
  4.1. W-based systems: dense film precursors......38
    4.1.1 Phase composition of the precursors......38
    4.1.2 Nanostructure, composition and thickness of the dealloyed films......42
    4.1.3 Time-dependent evolution of nanostructure and composition during dealloying......49

    4.1.4   Oxidation of the dealloyed films 58

    4.1.5   Photoelectrochemical properties of the dealloyed W-Fe dense films 59

  4.2   Ti-based systems: dense film precursors 64

  4.3   W-Ti-Mg ternary system: dense film precursors 68

  4.4   W-Fe system: nanocolumnar film precursors 71

    4.4.1   Nanostructure and composition of the nanocolumnar films 71

    4.4.2   Formation of nanostructures during dealloying 73

    4.4.3   Crystallographic characteristics of W-Fe nanocolumnar films 76

    4.4.4   Photoelectrochemical properties of the dealloyed W-Fe nanocolumnar films 77

5   **Conclusions and outlook** 79

6   **References** 81

List of abbreviations and symbols 89

1 Introduction
1.1 Solar-hydrogen technologies for sustainable energy supply

Nowadays clean and renewable energy supply is one of the biggest challenges for the mankind [1-3]. It becomes more and more important due to the exhaustion of natural reserves of fossil fuels [4], environmental pollution due to emissions of CO_2, SO_2, NO_x, etc. from hydrocarbon combustion by industries and auto transport, which leads to health problems among urban populations, global climate changes and other ecological problems [1, 3, 5]. Therefore, renewable and ecologically friendly sources of energy are considered as an alternative to fossil fuels. Hydrogen is often proposed as a prospective fuel of the future, but there are numerous scientific and technological problems to be solved on the way to the hydrogen economy [6]: starting from hydrogen production, storage, transportation, distribution and ending with efficient conversion of energy stored in H-H chemical bonds to electricity. Hydrogen is just an energy carrier and requires high amounts of energy for its production. Nowadays the cheapest industrial methods of hydrogen production are based on hydrocarbons decomposition: first of all steam reforming of natural gas [7] and partial oxidation of hydrocarbons [8]. But these methods do not give an advantage over direct use of hydrocarbons as fuels. The most sustainable way of hydrogen production would be splitting of water, as it is a part of the closed hydrogen cycle (Figure 1) and allows avoiding of undesirable by-products.

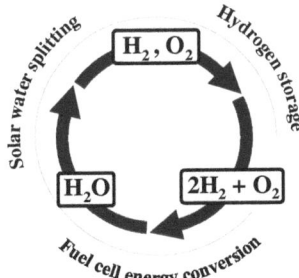

Figure 1: A scheme of the hydrogen cycle, including three stages: hydrogen production by water splitting, hydrogen storage (including transportation and distribution) and production of electricity by oxidation of hydrogen in fuel cells [9].

Direct water splitting by electrolysis requires enormous amounts of electric power, which can be generated either by nuclear energy or by renewable energies. Nuclear energy is not the best solution due to drawbacks with safety/security and problems with disposal of radioactive waste. So, one should look for a solution by implementing renewable sources of energy. Solar light is the main natural and renewable source of energy on Earth, but efficient conversion and storage of solar energy is still a problem [3, 10, 11]. One of the most promising approaches to convert solar energy is direct splitting of water into oxygen and hydrogen [10, 12, 13], because it is supposed to be a more cost efficient way of hydrogen production than usage of photovoltaic cells with further water electrolysis [11]. First, a direct process allows increase of the overall efficiency by cutting energy losses, unavoidable at intermediate steps. Additionally, by using a direct process, one can avoid expensive additional set-ups, which is essential for cutting technology costs [14]. Solar water splitting (also called photoelectrochemical water splitting [6] or artificial photosynthesis [15, 16]) is an electrochemical process of splitting H_2O molecules into H_2 and O_2 by passing electric current through an aqueous electrolyte, when the processes of electric current generation and electrochemical water splitting (electrolysis) are performed in one photoelectrochemical device (Figure 2).

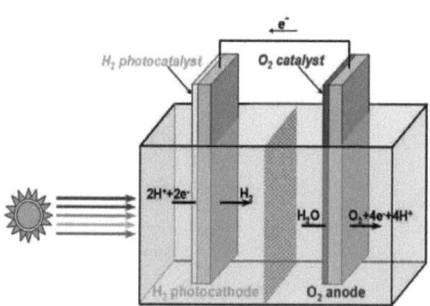

Figure 2: Schematic of a possible set-up for solar water splitting [17].

1.2 Materials for solar water splitting

Development of electrodes for solar water splitting is a materials science problem. One has to find the material with an appropriate band gap, band edge positions, high specific surface area and fulfill further requirements: long term chemical and mechanical stability, low cost of fabrication.

Figure 3 shows the characteristics of known and hypothetical materials for photoelectrochemical (PEC) water splitting. The minimal photovoltage must be higher than the thermodynamic value for the difference between the water oxidation and water reduction potentials (1.23 eV). Taking into account that the maximum obtainable photovoltage is usually equal to two thirds of the band gap of a good semiconductor material operating at its maximum power point, the band gap must be higher than 1.6 eV [18].

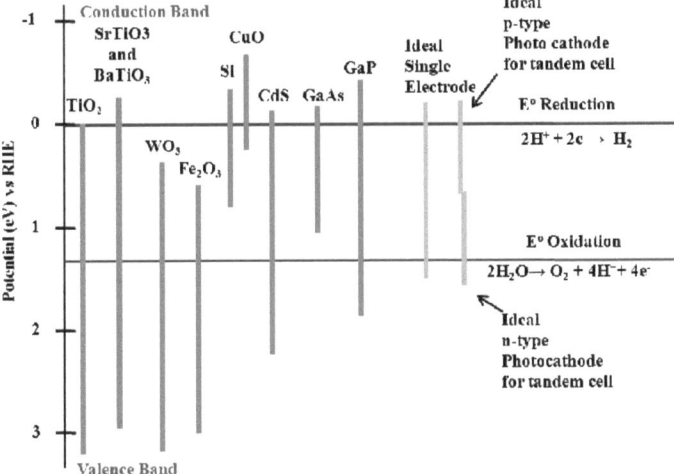

Figure 3: Band gap characteristics of known (green) and hypothetic (yellow) PEC materials [18].

An ideal single electrode should have a band gap of around 1.7 eV with a valence band edge more positive than the water oxidation potential and a conduction band edge more negative than the water reduction potential (Figure 4), so that the photogenerated holes and electrons have the driving force for the water splitting reaction [19]. If the PEC cell under illumination generates a potential <1.6 V, then an additional external electrical bias has to be applied (Figure 4d) to make the water splitting reaction possible. As an alternative, a combination of p-type and n-type semiconductor electrodes can be used. In this case the conduction band of the p-type photocathode material must be more negative than the water reduction potential and the n-type photoanode material must be more positive than the water oxidation potential. A photoanode and photocathode can be paired if the sum of their photovoltages at maximum power is greater than 1.6 eV. In addition the materials should have some catalytic activity for hydrogen or oxygen production in order to minimize the additional photopotential required to overcome the electrochemical overpotentials needed to drive the water oxidation and reduction reactions.

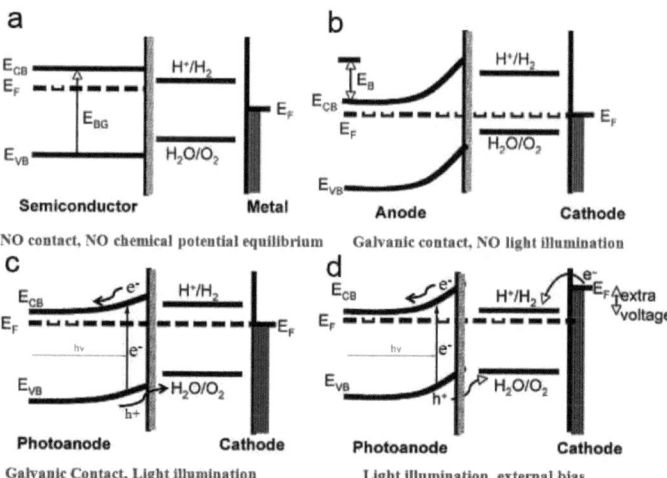

Figure 4: Semiconductor–electrolyte and metal–electrolyte junction energetics for the water splitting half-reactions in a PEC cell with semiconductor photoanode and metallic cathode under different conditions: a – no contact between electrodes and an electrolyte, no chemical potential equilibrium; b – galvanic contact between electrodes and an electrolyte established in dark conditions (without illumination); c – galvanic contact between electrodes and an electrolyte established under illumination; d – galvanic contact between electrodes and an electrolyte established under illumination and applied external electric bias [19].

During the last decades, three metal oxide based n-type semiconductors, TiO_2, Fe_2O_3, and WO_3, have been most extensively studied as potential candidates for photoanode materials in PEC cells [20-32] due to their good stability in aqueous electrolyte under continuous solar radiation and their relative abundance in nature. Despite the mentioned advantages, these materials still do not meet all the requirements. Anatase TiO_2 and monoclinic WO_3 possess good charge transfer properties but due to high band gaps (3.2 eV and 2.5–2.9 eV respectively) these materials absorb photons only in the UV range which is only a small part of the solar spectrum (Figure 5). This leads to low solar-to-hydrogen efficiency [33]. Hematite (α-Fe_2O_3) has an appropriate band gap (2–2.2 eV) to absorb in the visible range of the solar spectrum, but it possesses poor charge transfer properties and a very positive photocurrent onset potential (ca. 0.8 V vs. the standard hydrogen electrode, SHE), which requires the application of an additional electric bias in PEC water splitting cells [25]. In the case of WO_3 the photocurrent onset potential is much less positive, ca. 0.4 V vs. SHE [25]. This, together with the possibility to decrease the optical absorption edge of WO_3 by doping, makes WO_3 a promising material for PEC cells [34].

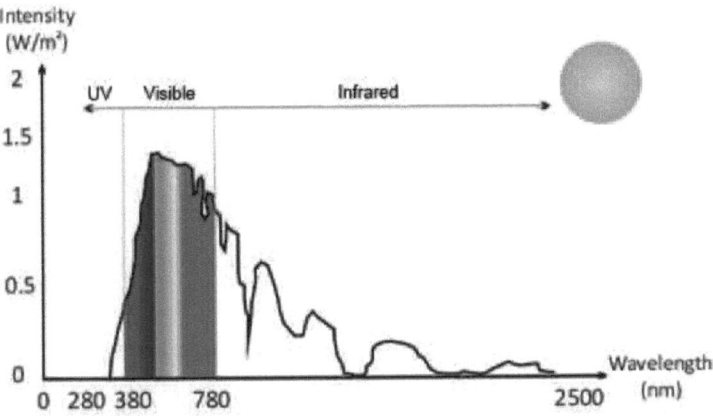

Figure 5: Solar spectrum at sea level [35].

The efficiency of oxide semiconductor photoanodes strongly depends on the morphology and thickness of the film [24, 25, 34, 36-38] and on the presence of doping elements in it [39-42]. A porous film structure leads to an increase of the overall efficiency of the photoelectrochemical (PEC) system because the photocurrent density is proportional to the specific surface of the electrodes. In addition, the porous structure enhances the photoconversion efficiency by rendering the path of photogenerated holes into the solution shorter and thus reducing electron-hole recombination. Furthermore, the porosity increases the light absorption, because light scattering inside a porous film is higher as compared to a dense film.

Apart from sol-gel methods [36], electrochemical anodization [37] and reactive magnetron sputtering [34], the dealloying process is one of the most promising methods for preparing high surface area electrode materials. Dealloying has been used mostly to synthesize porous metallic materials. Recently, it was used in combination with further thermal oxidation for the fabrication of porous oxides, in particular WO_3 [38, 43].

1.3 Aims of the thesis

The aim of this thesis is to develop porous thin film electrodes by a dealloying process and using combinatorial and high-throughput methods. These electrodes are intended to be used in the future in solar water splitting applications.

In the first two parts of the thesis the solar water splitting concept is described, the requirements of materials for solar water splitting are considered, fundamentals of the dealloying process as a fabrication method are analyzed, and the combinatorial materials science approach is introduced. In the third part the experimental methods used in this work are described. In the fourth and fifth parts the experimental results are described and discussed.

The following systems were investigated as possible precursors for the development of anodes for solar water splitting: W-Fe, W-Al, W-Mg, W-Au, Ti-Mg, Ti-Fe, and W-Ti-

Mg. These thin film materials libraries with continuous composition and thickness gradients were co-deposited by magnetron sputtering. The fabricated precursor materials were then chemically modified by dealloying with further thermal oxidation. Different dealloying conditions were compared. Electron microscopy (SEM and TEM), EDX, XRD and profilometry were used to characterize the materials. Finally, the functional properties of the porous electrodes were determined by PEC measurements.

2 Fundamentals
2.1 Chemical and electrochemical corrosion

According to the definition of the International Union of Pure and Applied Chemistry (IUPAC), corrosion is an "irreversible interfacial reaction of a material with its environment which results in consumption (e.g. dissolution) of the material or in dissolution into the material of a component of the environment" [44]. The mechanism of corrosion depends on the nature of the ambient environment and on electrical conductivity at the material-environment interface. Materials with high conductivity usually corrode electrochemically. At low conductivity, corrosion proceeds as a chemical process.

Conductive materials also corrode according to the chemical mechanism in environments with a lack of ionic conductivity, such as: crude oil and its derivatives, molten sulfur, a number of organic substances and some dry gases (H_2S, H_2, CO, CO_2, Cl_2, NH_3).

Conductive materials corrode electrochemically in environments of electrolytes. Conductivity depends on type of the media and composition of the material [45].

Electronic conductivity is typical for metals (conductivity, $1/\rho \sim 10^4$ $\Omega^{-1}cm^{-1}$), graphite ($1/\rho \sim 10^3$ $\Omega^{-1}cm^{-1}$), some carbides and nitrides of transition metals, some oxides and sulfides, e.g. FeS, PbS, CuS, Fe_3O_4, MnO_2. Conductivity for such compounds is in the range of 10 to 10^3 $\Omega^{-1}cm^{-1}$.

Ionic conductivity is typical for electrolytes, e.g. aqueous solutions of salts, acids and alkalis ($1/\rho < 1$ $\Omega^{-1}cm^{-1}$), molten salts ($1/\rho < 10$ $\Omega^{-1}cm^{-1}$), ionized gases, colloidal systems.

In most cases corrosion in aqueous solutions, called electrochemical corrosion, can be explained by the formation of corrosion microcells. The mechanism of formation of electric potentials on wet metal surfaces is similar to the one in galvanic cells [46, 47].

The following reaction takes place on the interface between metal (M) and solution during electrochemical corrosion in the anode area:

$$M \rightleftarrows M^{z+} + ze^- \tag{1}$$

In equilibrium conditions the redox potential of the half-cell can be described by the Nernst equation:

$$E = E^0 + \frac{RT}{zF}\ln\frac{a(M^{z+})}{a(M)} \tag{2}$$

Where: E^0 – the standard half-cell reduction potential; R – the universal gas constant: R = 8.314 472 J K^{-1} mol^{-1}; T – the absolute temperature; F – the Faraday constant, the number of Coulombs per mole of electrons: F = 9.648 533 99×10^4 C·mol^{-1}; z – the number of moles of electrons transferred in the half-reaction; a – the chemical activity for the relevant species, where a(M) is the reductant and a(M^{z+}) is the oxidant. $a_X = \gamma_X c_X$, where γ_X is the activity coefficient of species X. Since activity coefficients tend to unity at low concentrations, activities in the Nernst equation are frequently replaced by simple concentrations [M] and [M^{z+}].

[M] is usually a constant value for pure metals, while for alloys it is calculated from the chemical composition. Since the logarithm of a constant value is constant, the Nernst equation can be transformed to:

$$E = E_M^0 + \frac{RT}{zF}\ln [M^{z+}] \tag{3}$$

Where E_M^0 is the standard potential, determined empirically for each electrode reaction.

If the metal surface is not uniform, then several centers with different overpotentials for hydrogen ion reduction or dissolved oxygen reduction can be formed on wet metallic surfaces. This phenomenon leads to creation of local cathodes and anodes. The cathodic reaction is the reaction reduction of hydrogen ions:

$$2H^+(aq) + 2e^- \rightleftarrows H_2(g) \tag{4}$$

This process can take place only in the presence of an excess of electrons in the metal. The source of these electrons is the anodic process (1). Short-circuit of the cathode and the anode disturbs chemical equilibrium and leads to a flow of electrons, hydrogen gas formation, etc. Corrosion dynamics depend on the quantity of corrosion centers (microcells) and the properties of an electrolyte (temperature, concentration of the components dissolved in water, etc.).

The equilibrium electrode potential E is a relative value. Experimentally, only a potential difference between two electrodes can be determined. Thus one of these two electrodes is the reference electrode of known potential. The standard hydrogen electrode (SHE) is used as reference electrode and declared to be zero at all temperatures [44]. Potentials of any other electrodes are compared with that of the SHE at the same temperature.

The hydrogen electrode is based on the reversible redox reaction (4). This redox reaction occurs at a so-called platinized Pt electrode (bulk Pt electrode coated with a layer of highly dispersed Pt powder). The electrode is dipped in a solution of strong acid saturated with pure hydrogen gas under a pressure of 1 bar. The activity of both the reduced form (hydrogen gas) and oxidized form (hydrogen ions) are maintained equal to one $(a(H_2) = a(H^+) = 1)$.

Since the use of the SHE in measurements is impractical, other reference electrodes with a known potential vs. SHE are used. Frequently, these are:

1) Silver chloride electrode: Ag, $AgCl_{(s)}$/KCl (saturated); reaction: $AgCl(s) \leftrightarrow Ag^+ + Cl^-$; potential +0.22249 V (±0.13 mV) vs. SHE at T=298 K [48];

2) Saturated calomel electrode: Hg, Hg_2Cl_2 (calomel)/KCl (saturated); reaction: $Hg_2Cl_2 + 2e^- = 2Hg + 2Cl^-$; potential +0.2412 V vs. SHE at T=298 K [49];

3) Calomel electrode: Hg, Hg_2Cl_2 (calomel)/KCl (0.1 M); reaction: $Hg_2Cl_2 + 2e^- = 2Hg + 2Cl^-$; potential +0.3356 V vs. SHE at T=298 K [49];

The reference electrode is connected with the main reaction solution using an electrolytic bridge, eliminating the diffusion potential between solutions.

Equilibrium conditions for the electrode reaction can be represented graphically in the form of a Pourbaix diagram, indicating thermodynamically stable phases in dependence on the equilibrium electrode potential E and pH of a solution (Figure 6).

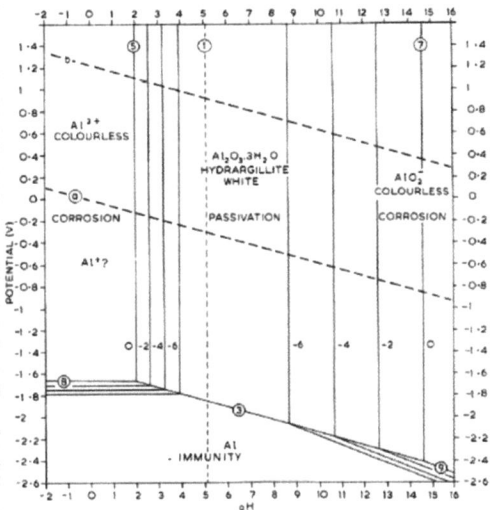

Figure 6: Potential-pH (Pourbaix) diagram for the Al-H$_2$O system at 25°C [50].

The exponent of the hydrogen ion concentration or pH of a solution is the negative logarithm of the concentration of H$^+$ ions:

(5)

The pH scale includes solutions from highly acidic to strongly alkaline.

The Pourbaix diagram contains regions where specific compounds or phases (solid Al, solid Al$_2$O$_3$, Al^{3+} and AlO$_2^-$ ions in aqueous solution in the Figure 6 example) are thermodynamically stable. The regions, where the dissolved ions are stable, are labeled "corrosion". The region of stable oxides and hydroxides is labeled "passivation", which means that metallic Al is protected with a film of oxide or hydroxide on the surface. The region where metallic Al is stable, is labeled "immunity".

Slanted dashed lines (a) and (b) in Figure 6 describe red-ox equilibrium conditions for water. Below the line (a) water reduction with gaseous hydrogen evolution is possible:

(6)

Above the line (b) water oxidation with gaseous oxygen evolution is possible:

$$2H_2O \rightleftarrows O_2 + 4H^+ + 4e^- \tag{7}$$

In between these lines, water is thermodynamically stable. The lines (a) and (b) can be described by the following dependences respectively:

$$E = -0.059pH \tag{8}$$

and

$$E = 1.23 - 0.059pH \tag{9}$$

Transitions of Al between different phases can be described by the following chemical equations:

$$Al \rightleftarrows Al^{3+} + 3e^- \tag{10}$$

$$2Al^{3+} + 3H_2O \rightleftarrows Al_2O_3 + 6H^+ \tag{11}$$

$$2Al + 3H_2O \rightleftarrows Al_2O_3 + 6H^+ + 6e^- \tag{12}$$

$$Al_2O_3 + H_2O \rightleftarrows 2AlO_2^- + 2H^+ \tag{13}$$

$$Al + 2H_2O \rightleftarrows AlO_2^- + 4H^+ + 3e^- \tag{14}$$

Similarly, equilibrium conditions of corrosion, immunity and passivation of Fe, Mg, W and Ti can be described by Pourbaix diagrams (Figures 7, 8, 9, 10).

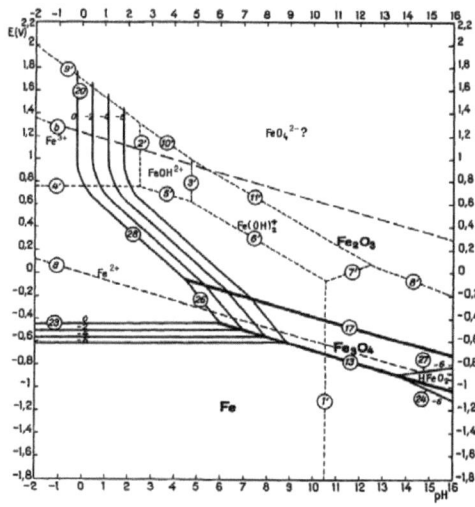

Figure 7: Potential-pH (Pourbaix) diagram for the Fe-H$_2$O system at 25°C [50].

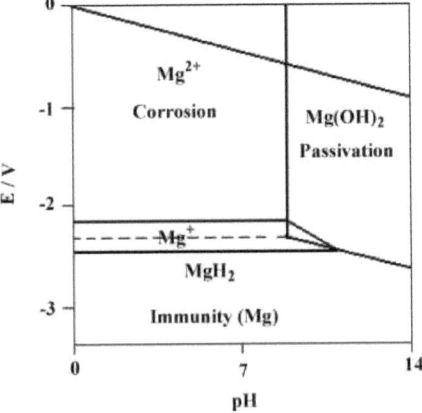

Figure 8: Potential-pH (Pourbaix) diagram for the Mg-H_2O system at 25°C [51].

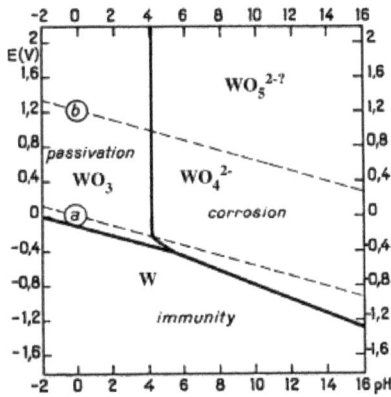

Figure 9: Potential-pH (Pourbaix) diagram for the W-H_2O system at 25°C [50].

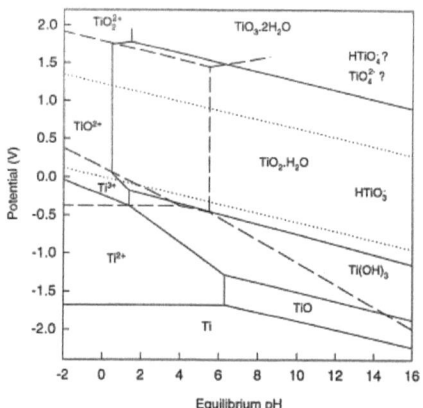

Figure 10: Potential-pH (Pourbaix) diagram for the Ti-H$_2$O system at 25°C [50].

Using Pourbaix diagrams can help to choose optimal conditions for metal etching. But one should take into account a number of limitations. Pourbaix diagrams contain information only about the thermodynamic stability of compounds but there is no information about kinetics. This means that certain compounds indicated in the diagram may not appear in real experiments when there is a very low rate of a reaction. Instead, thermodynamically unstable compounds may be observed. Another limitation of Pourbaix diagrams is that they are plotted for pure metals and simple solutions, not for alloys. Despite these limitations, the diagrams remain a powerful tool for electrochemistry and corrosion studies.

2.2 Selective corrosion of alloys

Selective corrosion, also called dealloying [43, 52, 53], is based on dissolution of the most electro-chemically active component in a two- or multi-component alloy due to the formation of local galvanic microcells in areas of local compositional microsegregation.

One of the components gets selectively dissolved. The resulting dealloyed material is porous and consists mostly of the more noble constituent of the initial alloy.

2.2.1 Kinetics of alloy corrosion

At a certain potential only the electrochemically more active component (A) will dissolve, while the less active component (B) will be thermodynamically stable. In this case all atoms A would be removed from the surface and atoms B would remain, and the dissolution process would stop. This does not happen, due to the surface mobility of the atoms of the more noble component (B) [54-56].

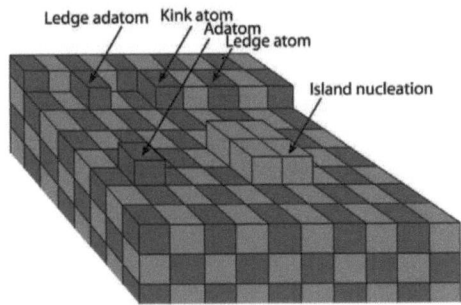

Figure 11: The scheme of homogenous alloy A–B: dark color – component A, light color – component B [57].

Even a well-polished surface of an alloy is relatively rough due to various defects. The simplest example of a surface defect is a ledge (Figure 11). Ledge and kink atoms possess higher energy because at least from two sides the neighboring atoms are missing. Adatoms possess even higher energy because they have neighbors only from one side. According to the surface diffusion model [54], during dealloying these atoms leave their original unstable positions and diffuse on the surface. Atoms A easily move to the etching solution when they have minimal number of neighbors. Remaining noble atoms B participate in nucleation on the surface and further growth of almost pure B crystals.

The surface diffusion mechanism was proposed by Gerischer [54] and was later improved by Forty [58] and Sieradzki [59]. Forty et al. proposed that both surface and volume diffusion are important for dealloying. Sieradzki et al. extended the surface diffusion model by percolation theory. Sieradzki et al. showed that selective dissolution is possible only above a threshold concentration of the more active element in the alloy [60]. The percolation model suggests that dealloying is possible only if the more electrochemically active element forms a continuous network of percolating clusters throughout the volume of an alloy (Figure 12) providing a path for dissolution. Below the critical concentration the network of percolation clusters becomes discontinuous which makes dealloying impossible.

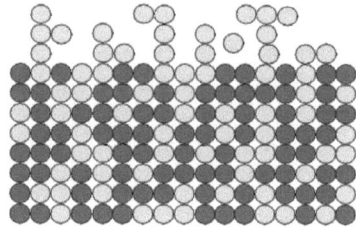

Figure 12: 2D scheme of percolation for alloy an A–B: gray color – electrochemically active component A, yellow color – component B. A network of interconnected clusters of B atoms is shown.

2.2.2 Thermodynamics of binary alloy corrosion

Thermodynamics of binary alloy corrosion should be considered to explain why the dissolution of one of the components in a binary alloy proceeds with higher rate, and sometimes the second component does not dissolve at all. As the process of alloy formation is spontaneous, the partial free Gibbs energies of the components decrease. In case of formation of an ideal solid solution, the free Gibbs energy change for one mole of the component "i" equals $\Delta G_i = RT \ln N_i$, where R – universal gas constant, T – absolute temperature and N_i – mole fraction of the component "i". This value is related to electric work $z_i F \Delta E_i$ by the equation $\Delta G_i = z_i F \Delta E_i$, where z_i – the amount of transferred electrons in the reaction $A \rightleftarrows A^{z+} + ze^-$; F – the Faraday constant and ΔE_i – the difference of equilibrium electrode potential of the component "i" in alloy and the pure component "i".

Therefore $\Delta E_i = -\frac{RT}{z_i F} \ln N_i$. (15)

As the molar fraction $N_i < 1$, therefore $\Delta E_i > 0$, i.e. partial potentials of the components are always more positive than for pure metals. This means that oxidation of metals from alloys will start at higher potentials [57]. The equation (15) also means that the rates of partial potentials change strongly depends on concentration of the component: at small N_i the change is big, at big N_i it is small [57].

Now one should consider, how the difference of standard electrode potentials of the metals A and B changes upon alloy (A-B) formation if $E^0_{A^{z+}/A^0} < E^0_{B^{z+}/B^0}$. Taking into account the equation (15), if $z_A = z_B = 1$ one can calculate that:

$$\Delta E_{A,B} = E_B - E_A = -\frac{RT}{F} \ln \frac{N_B}{N_A} \quad (16)$$

Thus, in case of an equimolar composition ($N_A = N_B = 0.5$) $\Delta E_{A,B} = 0$, i.e. the difference of electrode potentials of the components is the same as for pure metals. But the values of the potentials are positively shifted [57]. If $N_A \gg N_B$ then $\Delta E_{A,B} > 0$, i.e. the difference of properties of the components increases. If $N_A \ll N_B$ then $\Delta E_{A,B}$ decreases, and at a certain concentration one can reach an equality of electrode potentials. It is possible to estimate roughly this concentration. As $N_B \cong 1$, and $N_A \ll 1$, thus according to equation (15) electrode potential of component B will not change significantly. Thus, $\Delta E_{A,B} = -\frac{RT}{F} \ln N_A = -0.06 \lg N_A$. Now assuming that $\Delta E_{A,B} = E_B - E_A = 0.60\ V$, from the last equation one can calculate that $N_A = 10^{-10}$. This is a very small impurity concentration, which is quite rare for real alloys. It means that partial electrochemical properties of alloy components differ, thus selective dissolution of an alloy is almost always thermodynamically possible [57].

In most cases one of the components will be stable and will not dissolve if it has a higher standard electrode potential. For example in the Au-Ag alloy, which was extensively used for dealloying investigations [61-63], the difference of standard electrode potentials of the components (Au and Ag) is $\Delta E = E^0_{Au^{3+}/Au} - E^0_{Ag^+/Ag} = 0.8\,V$, which is a big driving force for selective Ag dissolution. In this case the required potential is being established due to formation of local galvanic cells. In case of some other alloys, an application of an external potential is required.

Figure 13 shows a sketch of anodic polarization curves for a binary alloy with different composition and for pure alloy components. The curves consist of two areas with growing current density separated by a plateau corresponding to a passivation region. The first increase in the current density with the potential is associated with dissolution of highly active kink atoms of the more active element A. The passivation of the electrode surface is related to depletion of the active atoms A on the surface and formation of a layer with the atoms B. To explain the passivation phenomenon different mechanisms were proposed [58, 64, 65] related to either surface diffusion of B or volume diffusion of A. Nevertheless, the reason of higher passive current density for alloys with higher amounts of the more active element is still unclear. At a certain potential (called critical potential, E_c) the current density rapidly rises due to the start of the less active metal dissolution and passive layer destruction. But the nature of the critical potential is still not explained, because even below the critical potential a pit formation was observed [66].

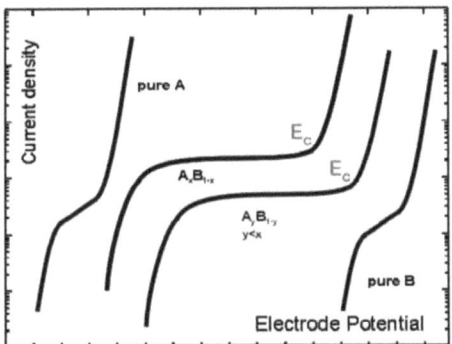

Figure 13: Schematic polarization curves of a binary alloy and pure alloy components during electrochemical dissolution [67].

2.3 Combinatorial materials science

A combinatorial materials science approach was first introduced by Boettcher et al. in 1955 [68]. The approach was implemented for rapid identification of metallic phases in ternary and quaternary systems and further construction of the phase diagrams. In 1965 the approach was again used for determining ternary alloy phase diagrams by Kennedy et al. [69]. Later in 70's combinatorial investigation of ceramic and metal oxide thin films was done by Hanak et al. [70, 71]. The combinatorial approach was shown to be very promising, but due to lack of high-throughput analytical methods and low speed of analysis the approach was not widespread for several decades.

The combinatorial materials science concept became more feasible in the 1990's due to rapid development of automated high-throughput screening techniques, growth in computational power and development of appropriate software for collecting data from the analysis instruments, processing and analysis of raw datasets with further visualization and distribution. It was successfully demonstrated by Xiang et al. in 1995 for the development of superconductors [72] and compounds with giant magnetoresistance [73]. The combinatorial approach was especially important for these groups of functional materials as they contain about six different chemical elements creating a lot of variables for a researcher.

The approach has been used for the development of functional materials: photocatalytic [74-78], photovoltaic [79-81], hydrogen storage [82-85], fuel cell electrode [86-91], Li-ion battery electrode [92, 93], shape-memory [94-100], optical [101-106], ferromagnetic [107-113], ferro- and dielectric [114-117], luminescence [118-125], organic polymer materials [126-128].

Using physical vapor deposition (PVD) processes one can produce thin film materials libraries with continuous composition and thickness spreads (Figure 14). This can be achieved by simultaneous sputtering of different materials with automatic mixing of the components during the deposition process. Composition and thickness spreads are determined by a certain positioning geometry of the deposition sources in relation to each other and to the substrate as well as by variation of the deposition rates.

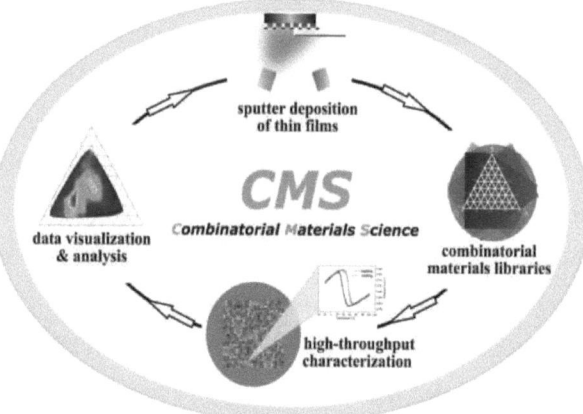

Figure 14: A visualization of combinatorial materials science and high-throughput experimentation methodology [129].

For materials libraries characterization, various high-throughput methods are applied. Extensively used methods for screening composition and structure of thin film materials libraries are commercially available X-ray diffraction (XRD) and energy dispersive X-ray analysis (EDX). During the last years these methods were significantly improved in terms of spatial resolution and measurement times. Film thickness measurements can be performed by high-throughput profilometry measuring the steps between a substrate and film surface, created by a photoresist lift-off mask.

For characterization of specific functional properties of materials several techniques were developed: scanning droplet cell for electrochemical [130] and photoelectrochemical [38] high-throughput analysis, optical analysis of hydrogen storage materials [83], scanning squid microscopy [131], methods for screening thermoelectric properties [132] etc.

3 Experimental methods

3.1 Preparation of thin film materials libraries

The search for an appropriate material with the best PEC properties requires a high number of experiments in order to find optimal nanostructure (providing high specific surface and stability of the electrodes), film thickness and composition. Therefore, in this thesis the combinatorial materials science approach and high-throughput techniques [129, 133] were used for screening through a high number of compositions and structures. The fabrication technique of porous oxide thin-film electrodes consists of the following steps (Figure 15): combinatorial sputter deposition of metallic alloys, subsequent chemical selective dissolution (dealloying) of the thin film materials libraries, and finally oxidation of the metallic porous films by annealing in air. Below all of these steps will be described in more details.

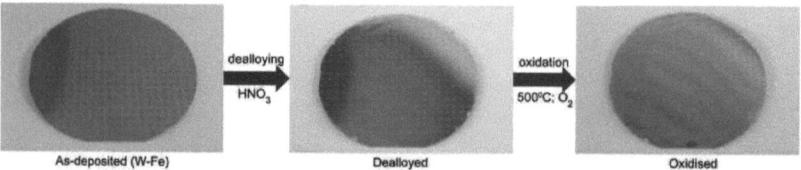

Figure 15: Materials libraries fabrication steps: as deposited, dealloyed and subsequently oxidized thin film materials library deposited on Si(100) wafers (4 inch diameter) with 1.5 µm thick thermal SiO_2 layer.

3.1.1 Magnetron sputter deposition

Precursor W-Fe, W-Al, W-Mg, Ti-Mg and W-Ti-Mg thin film materials libraries with continuous composition and thickness spreads were co-deposited by magnetron sputtering on thermally oxidized Si(100) wafers[1]. Metallic W, Ti, Fe, Al and Mg targets (99.99% purity, 4 inch diameter) were used in a magnetron sputter system (AJA Interna-

---

[1] Si(100) wafers (4-inch diameter) with a 1.5 µm thick thermally-grown continuous SiO_2 layer serving as a diffusion barrier.

tional) (Figure 16). The sputter system consists of a load-lock and a sputter chamber. The substrates are transferable from the load-lock to the sputter chamber and back by a mechanical manipulator. The sputter chamber is equipped with 4 magnetron cathodes (connectable to direct current (DC), pulsed direct current (pDC) or radio frequency (RF) power supplies) mounted above a rotatable substrate holder with an integrated heater. The typical base pressure before sputtering was $\leq 8.5 \times 10^{-6}$ Pa. The targets were pre-sputtered against closed shutters, in order to minimize contamination from absorbed substances and surface oxides. All films were deposited in Ar atmosphere at room temperature, i.e. without intentional heating; the sputter pressure was 1.3 Pa. Thermally oxidized 4-inch Si(100) wafers[1] were used as substrates. For enabling photocurrent measurements, a conductive 100 nm thick Pt layer was deposited by magnetron sputtering prior to the fabrication of the thin film materials libraries. First, thin 10 nm layer of Ti was deposited (DC magnetron sputtering with a sputter rate of 0.27 nm/s) to improve adhesion of the Pt film to the substrate. Then, the Pt layer was deposited by DC magnetron sputtering with a sputter rate of 0.1 nm/s. The substrates were continuously rotated during sputtering to provide uniform film thickness.

Figure 16: Combinatorial magnetron sputter system (AJA International).

For the fabrication of *dense film* materials libraries the targets were located at a 90° angle between each other and were inclined at an angle of 26° with respect to the substrate normal (Figure 17). This deposition geometry allows to produce materials libraries with a thickness gradient (120 nm–300 nm) as well as a composition gradient (W_x(soluble element)$_{1-x}$, Ti_xMg_{1-x} ($0.9 < x < 0.35$) and $W_xTi_yMg_{1-x-y}$ ($0.6 < x < 0.1$; $0.3 > y > 0.8$)). The sputter rate was 0.16 nm/s for W, 0.27 nm/s for Ti, 0.05 nm/s for Fe, 0.05 nm/s for Al, and 0.1 nm/s for Mg.

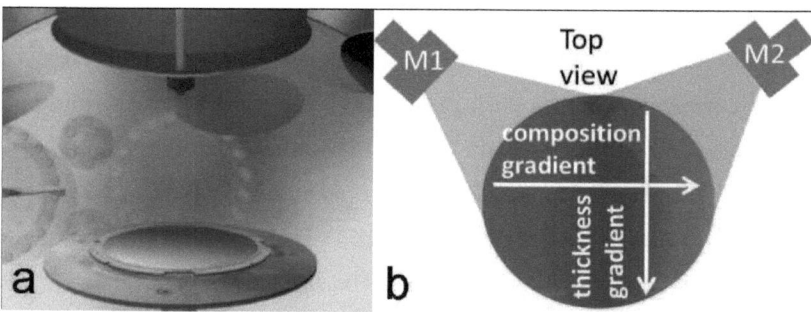

Figure 17: Schematic of the fabrication of materials libraries with continuous composition and thickness gradients by magnetron co-sputtering. (a) Schematic internal view of the chamber during the co-sputtering process[2]. (b) Schematic top view[3] of a binary materials library depicting composition and thickness gradients formation when the targets are located at 90° angle between each other.

In order to grow W-Fe *nanocolumnar films* the glancing angle deposition (GLAD) technique [134-136] was used[4]. This technique is based on a self-shadowing mechanism. A flow of atoms from the gas phase impinging on a substrate from an oblique angle in vacuum leads to growth of columnar structures. The angle of incidence affects the degree of shadowing and thus influences the morphology. Rotation of the substrate is another important factor influencing the column shape. The W target was attached with an oblique inclination angle of ~ 88°, while the Fe target was positioned at an inclination

---
[2] Drawn by Dario Grochla.
[3] Drawn by Martin Hofmann.
[4] Dr. Chinmay Khare is specially acknowledged for the help with GLAD depositions.

angle of ~ 27°, with respect to the substrate normal, as schematically shown in Figure 18. In order to determine the effective deposition rates, a mechanical profilometer (Ambios) was used, which measured the step between substrate and film surface, created by a photoresist lift-off mask. The deposition rates were then adjusted (W: 3.5 nm/min and Fe: 0.66 nm/min) in order to grow W-Fe nanocolumns. W and Fe targets were sputtered using 300 W DC and 50 W RF, respectively. In order to grow vertical nanocolumns, substrates were continuously rotated with 25 rpm. During the deposition process an initial layer of pure W was deposited (~ 100 nm) and subsequently with co-deposition of W and Fe for t_d = 125 min, W-Fe nanocolumns were grown thereon. Figure 18b shows a schematic of the as-grown nanocolumns. A reference GLAD pure W nanocolumnar film (thickness ~ 450 nm) was additionally deposited and examined in order to compare the film characteristics.

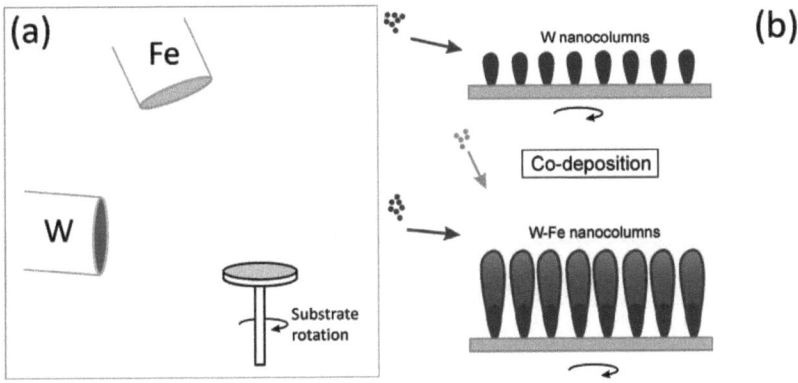

Figure 18: Schematic of the fabrication process of nanocolumnar films by glancing angle deposition (GLAD)[5]: (a) Schematic side view of the sputtering chamber configuration for nanocolumnar films deposition. (b) Schematic view of as-grown nanocolumns.

---

[5] Drawn by Dr. Chinmay Khare.

3.1.2 Dealloying

For dealloying, the as-deposited thin film materials libraries were immersed in aqueous HNO_3 solutions with different concentrations (0.12 wt.%, 1.0 wt.%, 3.5 wt.% and 10 wt.%). This process was performed at room temperature under open circuit conditions for different times (from 5 minutes to 48 hours). In acidic solutions, the standard electrode potentials of the W^{6+}/W, Ti^{2+}/Ti, Fe^{3+}/Fe, Al^{3+}/Al and Mg^{2+}/Mg couples are 0.680 V (standard hydrogen electrode, SHE), −1.630 V, −0.037 V, −1.670 V and −2.363 V (SHE), respectively. W is thermodynamically more stable than Fe by 0.717 V, Al by 2.350 V and Mg by 3.043 V. Ti is thermodynamically more stable than Mg by 0.733 V. Hence, during the dealloying process Fe, Al or Mg atoms are selectively dissolved. The remaining W and Ti self-assembles to crystalline W and Ti, respectively, with an altered nanostructure. To identify roughly at which pH and external potential selective dissolution of an alloy is possible. Pourbaix diagrams (Figures 7-10) can be used. Pourbaix diagrams are built for pure metals, not for alloys, but comparison of the thermodynamic areas of immunity/passivation/corrosion for each component of an alloy can give an estimation of possible dealloying conditions. After dealloying, the materials libraries were rinsed with deionized water and blow-dried in warm air. To identify the influence of crystallinity and presence of intermetallic phases in a precursor on dealloying, one precursor W-Fe library was annealed at 950 °C for 10 h in vacuum (p ~10^{-7} Pa) to induce phase formation, including the intermetallic phase Fe_7W_6. Finally, the dealloyed films were transformed into oxides by annealing at 500 °C in air for 8 h with heating and cooling rates of 2 K/min. Higher annealing temperatures are inappropriate because of degradation of the nanostructure due to intensive diffusion.

3.2 Characterization

3.2.1 Compositional analysis – energy dispersive X-ray analysis (EDX)

Energy dispersive X-ray (EDX) analysis allows the determination of the elemental composition of a sample. A high-energy focused electron beam excites electrons from an inner shell of an atom. The excited electrons can be ejected creating a hole. Another electron from a higher energetic level fills the hole and emits characteristic X-ray radiation corresponding to the difference of the energetic levels. The structure of energetic levels is specific for every atom, which allows determination of the chemical elements by measuring the emitted characteristic X-ray radiation.

EDX analysis (INCA x-act OXFORD instruments) was used for high-throughput determination of the element composition of the thin-film materials libraries. A cobalt standard was used for calibration of the instrument prior to each measurement. This allows achieving an accuracy of ±0.5 at.% using a measuring time of around 100 s per measurement point. A measurement grid consisting of 342 points with an x and y spacing of 4.5 mm was defined. The high-throughput mode allows scanning of the complete materials library automatically, and obtaining a text file with information about the composition in every measurement point with respect to the x, y coordinates. This text file is used as the data source for further creation of color-coded composition maps using Origin scripts.

3.2.2 Structural analysis – X-ray diffraction and X-ray scattering techniques

High-throughput X-ray diffraction (XRD) analysis (X'Pert PRO X-ray diffractometer PANalytical) was used to identify the crystalline phases in as-deposited, dealloyed and oxidized films. The system was calibrated using a Si standard.

The principle of X-ray diffraction analysis is the following: an X-ray beam generated usually by a Cu source is diffracted at atomic lattices according to Bragg's law (17), and the intensity of the diffracted beams is measured by an X-ray detector.

$$n\lambda = 2d \cdot sin\theta \tag{17}$$

where n – an integer number, λ – wavelength of the incident x-ray wave (0.154 nm for Cu K_α source), d – inter-planar lattice distance, θ – the angle between the incident X-ray and the scattering atomic planes.

The spacing between the planes in the atomic lattices (d) and diffraction peaks intensities can be used for identification of crystalline phases. The identification of the detected phases was performed by comparison of diffraction patterns to databases of known phases: inorganic crystal structures database (ICSD) and Pauling File binaries.

In the present work, the Bragg-Brentano geometry, the most common working mode of XRD, was used. In this mode both the X-ray source and the detector are symmetrically moved in such a way that an angle between the incidence and the diffracted beams equals 2θ. The Bragg-Brentano geometry is used for phase analysis of relatively thick films (usually with thickness above 100 nm).

For phase analysis of thin layers, small angle X-ray diffraction (Grazing-Incidence XRD) could be used. In this case the X-ray source is fixed at a low angle (<5°) in respect to the sample surface to avoid penetration of X-rays in depth of the sample, and the scan is performed only with the X-ray detector.

At certain conditions grazing incidence geometry (grazing incidence small-angle X-ray scattering, GISAXS) could be used to quantify film porosity: it can measure the specific inner surface area, average pore size and pore size distributions. X-ray scattering is different from X-ray diffraction and based on detection of the elastically scattered X-rays and measuring their intensity in dependence on incident and scattered angles, polarization, and wavelength. GISAXS is a special technique for measuring porosity in thin films, because the conventional N_2 gas adsorption technique which is used for measuring porosity of powders is not applicable for thin films due to the very small amount of materials in the thin layers. However GISAXS is not applicable for the films with high surface roughness.

3.2.3 Film thickness measurements

Profilometry (Ambios) was used for high-throughput film thickness measurements. A grid of photoresist stripes was prepared on the wafer surface before film deposition. After the films were deposited, the photo resist was removed by treating the wafers sequentially with different organic solvents (*N*-Methyl-2-pyrrolidone, acetone and isopropanol) in an ultrasonic bath. Film thickness was determined by measuring the height of the steps between a film and a substrate with a stylus profilometer (Ambios).

Another method – scanning electron microscopy (SEM) was also used for film thickness determination at film cross-sections prepared by mechanical cracking or cut by using focused ion beam (FIB). This method is more time consuming but it allows measurement of the thickness of each layer in a multilayer film.

3.2.4 Scanning electron microscopy (SEM)

Scanning electron microscopy (SEM) by Leo 1530VP was used for characterization of the film morphology, determination of grain shapes and sizes, and qualitative comparison of porosity in different samples. Cross-sectional SEM images were made to determine how the porous structure changes from the film surface to the substrate and to measure the thickness of layers.

SEM is based on scanning the sample surface with high-energy focused electron beam. An electron beam is thermionically emitted from an electron gun (usually with a W filament cathode) and focused with a number of electromagnetic lenses and metallic apertures. The primary electron beam usually has an energy ranging from 200 eV to 40 keV. When the electron beam hits the sample, electrons interact with atoms in a certain droplet-shaped volume. The penetration depth and the size of the interaction volume depend on the energy of the electron beam, atomic numbers of the specimen and the density of the material. Low energy (<50 eV) secondary electrons emitted by inelastic scattering are coming from the surface area (up to 2 nm in depth). High energy inelastically backscattered electrons are emitted from depths up to 0.1-5 µm (depending on the interaction volume size). The intensity of the secondary electrons beam is very sensitive to the incidence angle, which makes secondary electrons very suitable for sample topography visualization. The intensity of the backscattered electrons is sensitive to specimen composition, because heavy elements scatter electrons stronger than light elements, which allows observation of observe phase contrast.

3.2.5 Transmission electron microscope (TEM)

TEM was used for morphology and structure characterization with high spatial resolution[6]. For TEM analysis, a thin film lamella was prepared by FIB. Additionally, free standing nanostructures from the dealloyed film with the most prominent nanoblade-like structures was prepared on a specimen holder (3 mm diameter disk), with an electron beam transparent window made of amorphous Si-N membrane. TEM micrographs were obtained in a FEI Technai F20 G2 TEM at an acceleration voltage of 200 kV.

---

[6] Dr. Pio John Buenconsejo is specially acknowledged for the help with TEM measurements.

TEM is based on transmission of an electron beam through an ultra-thin sample and interaction of electrons with the sample when passing through it. The transmitted beam contains information about electron density of the sample, phase and periodicity, which are used for image formation.

3.2.6 Photoelectrochemical (PEC) measurements

Assuming that the origin of the photocurrent is the water splitting process, the value of the photocurrent can be used as a first approximation of the conversion efficiency [137]. Taking into account convenience of amperometry in comparison to detection of gaseous hydrogen or oxygen (the products of water splitting), photocurrent was used as the screening parameter for characterization of PEC activity of the electrodes. PEC characterization by means of a robotic setup is applicable for high-throughput screening of the materials libraries. High-throughput PEC screening helps to find an area with optimal properties (high conversion efficiency and corrosion stability) within a materials library.
A new PEC screening set-up based on the combination of a droplet cell and an automated robotic system was developed by Kirill Sliozberg[7] in the group of Prof. Dr. Wolfgang Schuhmann. A light fiber was introduced in a conventional droplet cell thus it represents a real three-electrode arrangement with light source. The system is able to perform all kinds of electrochemical and photoelectrochemical measurements on small spots of large materials libraries. The cell was equipped with a new kind of very durable tip, which allows performing of thousands of highly reproducible measurements and simultaneously solves the technical problem of the light fiber placement and homogenous sample illumination.
The following parts were used to assemble the set-up: the steper motor-driven stages LTM 60 series and stepper motor card PS-30 were purchased from Owis (Staufen, Ger-

---

[7] Kirill Sliozberg is specially acknowledged for the help with PEC measurements.

many); a force sensor KD 45-2N was bought from ME-Meßsysteme GmbH (Hennigsdorf, Germany); Xe-Lamp LC8 was made by Hamamatsu Photonics GmbH (Herrsching am Ammersee, Germany); power meter NOVA II with an in-house modified measuring head 3A-SH was purchased from Ophir Optronics Ltd. (Jerusalem, Israel); the 16-bit AD/DA card CIO-DAS1602/16 was obtained from PlugIn Electronic (Eichenau, Germany); a Jaissle potentiostat 1002 PC.T. was used for electrochemical experiments; syringe pump Cavro XLP 6000 was purchased from Tecan (Männerdord, Germany). The droplet cell and additional parts were made by the mechanical workshop of the Ruhr-University Bochum. Metal wires for miniaturized Pt counter-electrodes and Ag/AgCl reference electrodes were purchased from Goodfellow (Bad Nauheim, Germany). The software for data acquisition and control of the hardware components was developed using Visual Basic 6.0.

Automation of the screening process was realized by using robotic systems. Such systems were extensively used in the group of Prof. Schuhmann in the field of bioelectrochemistry for i.e. redox modification of proteins [138], combinatorial preparation and evaluation of biosensors [139] as well as for development and optimization of redox catalysts for voltammetric nitric oxide sensors [140] or testing of heterogeneous catalyst libraries [141]. Development and adaptation of electrochemical robotic systems to practical applications depending on desired application was described [142]. Photoelectrochemical measurements can be performed with such a system in which a light source is coupled in using a dedicated optic fiber. One particular technique that is commonly employed in screening of large compound libraries is the use of microtiter plates. Since screening of small spots of solid-state materials is considered, another approach has to be found. In an attempt to open the full spectra of electrochemical methods for application in a confined area, the scanning droplet cell was invented [130]. The basic setup makes use of a capillary that is wetting the surface of interest either with a free hanging droplet or using a silicone seal at its tip. A reference microelectrode is built into this capillary and wrapped or accompanied by a counter electrode. Thus, spatially resolved surface analysis becomes possible.

Since most metal oxides are hydrophilic and a free droplet requires non-wetting surfaces (contact angles close to 90°), it was necessary to use a capillary with a silicone rubber gasket [143]. Instead of a glass capillary with silicone rubber as sealant, a specially designed polytetrafluoroethylene (PTFE) capillary was used. In addition to its chemical stability, such factors as hydrophobicity and plasticity of PTFE were the reason to choose this material for this particular application. The realization of this tip is a better alternative to capillaries with silicone rubber gaskets in terms of robustness and allows easier technical solutions for light fiber placement. The designed optical scanning droplet cell (OSDC) (Figure 19) consists of a body made of PMMA (1), which incorporates an optical fiber (2), electrolyte inlet (3), capillary tip (4), miniaturized Ag/AgCl reference electrode (5) and Pt counter electrode (6).

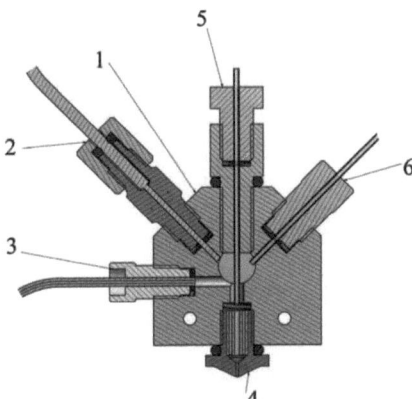

Figure 19: Scheme of the OSDC[8]: PMMA body (1), optical fiber (2), electrolyte inlet (3), PTFE tip (4), miniaturized Ag/AgCl reference electrode (5), Pt - counter electrode (6).

A special centric adapter (2) inside of the tip body (1) was used for stationing the light fiber (4) exactly against the nozzle (Figure 20). An O-ring (3) is used to seal the tip-cell

---

[8] Drawn by Kirill Sliozberg

connection. The openings in the centric adapter are machined for electrolyte flow and for the Pt-wire as the counter electrode (5).

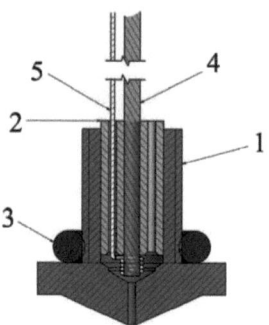

Figure 20: Scheme of the tip[9]: PTFE body (1), PEEK centric adapter (2), O-ring (3), light fiber (4), Pt - counter electrode (5).

The used light fiber had a core diameter of 1 mm and was made of PMMA. It is transparent for the wavelength range of 300 nm to 800 nm. Edge effects due to inhomogeneous illumination were avoided by designing the opening diameter to be smaller than the diameter of the light fiber. In this way, the entire active area of the sample was uniformly illuminated.

In order to avoid leakage of the electrolyte during the measurements, the front plane of the PTFE tip was polished using lubricated 1 μm sandpaper to get the tip surface smooth and exactly parallel to the sample plane. The wetted area of the sample surface formed the working electrode. This was reached by pressing the tip slightly against the sample, which was then moistened with the electrolyte. Due to the plasticity of PTFE, the walls of the tip are slightly deformed. A hermetic seal between the tip and the sample is thus established. It is important that the applied pressure does not lead to inelastic deformation of the tip, otherwise the surface area is not reproducible and electrolyte leakage

---

[9] Drawn by Kirill Sliozberg

cannot be avoided. Typical applied pressures varied from 100 mN to 600 mN depending on the tip size and geometry. The diameter of the tip nozzle can be varied over a wide range depending on the application. For the PEC measurements, a tip with a nozzle diameter of ca. 0.82 mm was used. The geometrical wetted area of the sample was the same as that of the nozzle and was fairly reproducible independently of surface properties. Since electrochemistry occurs at an electrode/electrolyte interface, only the portion of the sample that is in contact with the electrolyte is probed. This allows analysis of small spots within a large library of materials, which possess property gradients. The resistivity of the cell was estimated to be 400 Ohm on Pt work electrode using 0.5 M $HClO_4$ solution as electrolyte. This resistance value is acceptable, since measured currents are typically in the region of tenths of µA to a few µA.

Figure 21 shows the main components of the set-up. The flat materials library to be investigated can be moved in the X-Y-plane by stepper motors. The OSDC is moved vertically (Z-axis) and a force sensor is used for controlled approach of the cell to the sample surface. The sensor is connected to the combined amplifier/ADC-unit, which transfers the measuring signal to the PC via a USB port. A Xe lamp with wavelength range of 400 nm to 700 nm is used as the light source. It is connected to the cell with a thin PMMA light fiber. The built-in shutter is controlled via serial port. The system uses a syringe pump to force the electrolyte through the body of the droplet cell to the tip. An analog potentiostat is connected to the computer by means of an AD-DA card.

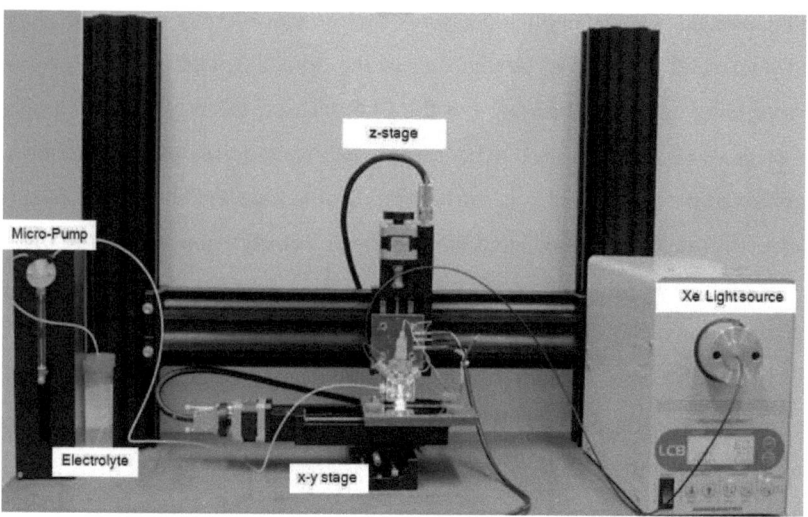

Figure 21: Setup for PEC measurements (in the group of Prof. Dr. W.Schuhmann).

The set-up was operated in stop-and-go mode, that is, the cell moves to the first position, collects local data and then moves to a new position. In order to exclude contamination and to avoid degradation of the electrolyte, the cell was moved to the waste position and rinsed with fresh electrolyte after each single measurement. In order to get the same droplet size after each rinsing procedure, a special wiper was used to remove any residual electrolyte droplets from the nozzle plane. After the rinsing procedure, the OSDC was prompted to move above the wiper. Polystyrene was chosen as the wiping material because of its hydrophilicity, which is slightly higher than that of PTFE, thus the eventually remaining drop is taken off by the wiper.

Since the assembly corresponds to the conventional 3-electrode arrangement with the sample acting as the working electrode, all the common potentiostatic and galvanostatic techniques, including impedance spectroscopy, cyclic voltammetry, or simple/multistep amperometrical detection, are possible. For photocurrent detection the current was recorded at 1 V bias potential at different positions. The applied potential was measured against the miniaturized built-in Ag/AgCl/3 M KCl reference electrode. 0.5 M $NaClO_4$

solution with pH 4.5 was used as the electrolyte. The amperometrical detection was employed because it allows detection of steady state currents. This is useful information, since real devices for solar water splitting will be most likely operated in maximum current mode. The simplest measuring procedure at every point includes rinsing of the cell with fresh electrolyte, positioning of the cell, measurement of the dark current and measurement of the current upon illumination of the sample. The measured photocurrent was calculated as the difference between the stabilized current recorded with the light switched on and the current recorded in the dark (Figure 22).

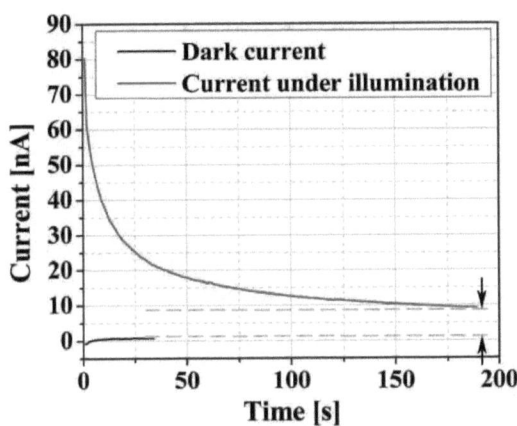

Figure 22: Photocurrent measurements.

The incident light power was set to 100 mW/cm^2. This value corresponds to the value of the AM1 irradiation, which is about 100 mW/cm^2 if the sun is at the zenith. The current detection was done not for a fixed duration, but timed to be long enough that a steady current value was reached. The current was considered as stabilized if its average value did not change by more than 1 % within 10 seconds. For evaluation, the current values of the last 10 seconds are averaged. All reported currents are normalized. Scanning of the surface was performed point-by-point, leading to generation of a data map.

4 Results and discussion

As shown in chapter 1.2, WO_3 and TiO_2 were chosen as the most promising materials for photoanode development. Porous oxide thin film materials libraries were fabricated according to the general procedure described in chapter 3.1: sputtering of W- and Ti-based metallic alloys followed by chemical dealloying with further thermal oxidation. The precursors for dealloying (W-Fe, W-Al, W-Mg, Ti-Mg and W-Ti-Mg) were chosen according to the requirements described in chapter 3.1.2.

The "Results and discussion" section consists of 4 chapters (with subchapters): 4.1 "W-based systems: dense film precursors", 4.2 "Ti-based systems: dense film precursors", 4.3 "W-Ti-Mg ternary system: dense film precursors", 4.4 "W-Fe system: nanocolumnar film precursors". Chapters 4.1, 4.2, and 4.3 are devoted to the fabrication of photoanodes by dealloying of <u>dense</u> film precursors (W-, Ti-, and W-Ti-based alloys respectively). Chapter 4.4 is devoted to a more advanced fabrication technique, where the precursor film is not dense but <u>nanocolumnar</u>. The advantages of this approach are discussed there.

4.1. W-based systems: dense film precursors
4.1.1 Phase composition of the precursors

XRD analysis was performed at every stage of sample preparation in order to control phase composition of the films. Figures 23-25 shows a series of XRD patterns of the as-deposited thin film materials libraries W-Fe, W-Al and W-Mg.

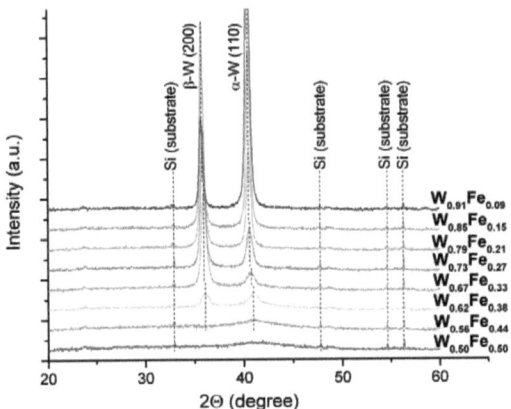

Figure 23: Color-coded XRD patterns from the as-deposited W-Fe materials library for measurement regions with different W:Fe ratios.

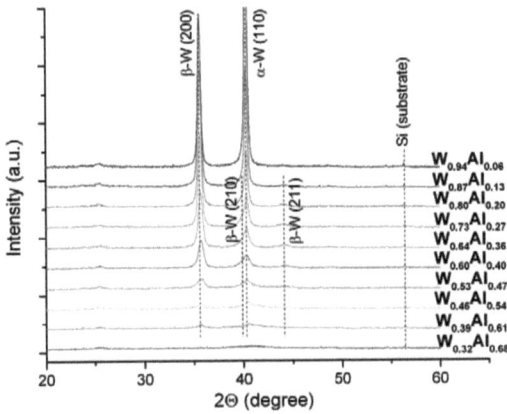

Figure 24: Color-coded XRD patterns from the as-deposited W-Al materials library for measurement regions with different W:Al ratios.

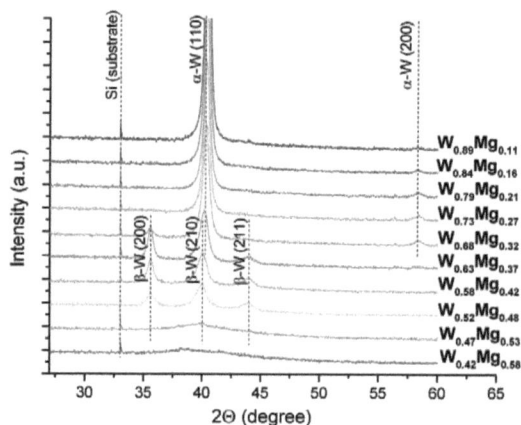

Figure 25: Color-coded XRD patterns from the as-deposited W-Mg materials library for measurement regions with different W:Mg ratios.

The peaks corresponding to the α-W and metastable β-W phases are observed for most compositions. The thermodynamically stable α-W phase with a lattice parameter of 0.316 nm dominates at compositions with low content of Fe, Al and Mg. With increasing content of these elements in the alloys the metastable β-W phase with increased lattice parameter (0.505 nm) becomes predominant due to the stabilization effect [144] in solid solutions. It is known, that the presence of other elements in interstitial positions of the W lattice could destabilize the lattice cells mechanically, which could prevent the growth of the α-W crystallites [145]. The peaks matching to Mg, Al and Fe are missing because W forms solid solutions with Mg, Al and Fe with the crystalline structure of cubic W in a broad range of concentrations. This assumption is reasonable because formation of solid solutions with the crystalline structure of bcc W by substituting W atoms with other elements is a common property of many W-based alloys [145]. An increase in the lattice parameter of the W phases with the increasing content of the alloying element is also clearly visible from the XRD patterns by peak shifts. Since the atomic radius of Fe is lower than that of W, the substitution of W by Fe leads to contraction of the lattice.

By contrast, the atomic radii of Al and Mg are higher than that of W, which leads to dilatation of the lattice. In addition, the peak intensities sharply decrease with decreasing W concentration, indicating the decrease of crystallinity. Compositions with ≤50 at.% W appear amorphous in XRD, because high concentrations of Fe, Al and Mg hinder W crystallization due to the big difference in atomic radii [146] and non-equilibrium film growth conditions [146, 147]. After annealing at 950°C in vacuum, metastable β-W phase undergoes transformation to the thermodynamically stable α-W phase. At the same time in the case of Fe-rich compositions (62 at.%), the formation of the intermetallic compound Fe_7W_6 was observed (Figure 26).

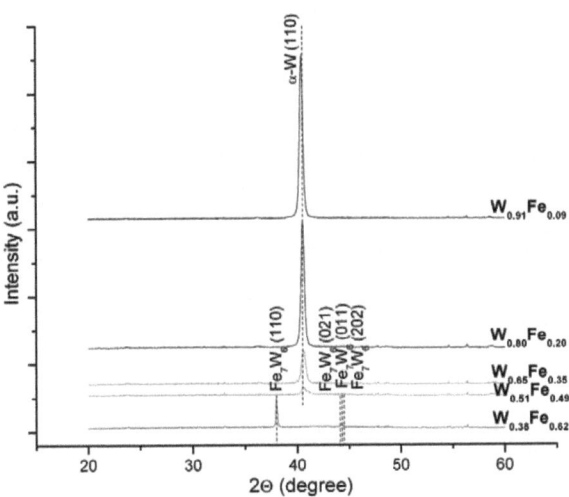

Figure 26: Color-coded XRD patterns of the annealed W-Fe materials library (950°C, 10 h) with different W:Fe ratios.

4.1.2 Nanostructure, composition and thickness of the dealloyed films

Figure 27 shows SEM images of the as-deposited $W_{0.7}Al_{0.3}$ and $W_{0.7}Fe_{0.3}$ thin films which were dealloyed for 24 h in aqueous HNO_3 solutions of different concentrations. These compositions were chosen as examples, because the shapes and sizes of nanocrystals after dealloying at certain etchant concentration look similar for all compositions of the precursor. The only difference for films produced from precursors with high contents of soluble metal is that the films become looser and discontinuous. However, the images demonstrate that there is a strong dependence of the nanostructure, i.e. the shapes and sizes of the nanocrystals of the dealloyed films, on the concentration of the etching solution.

Dealloyed films consisting of cross-linked nanoflakes were obtained when etching the precursor films in acid solutions with concentrations of HNO_3 3-10 wt. %. In addition, the length of the nanoflakes increase with increasing acid concentration (e.g. from $30\times300\times300$ nm^3 at 3.5 wt. % HNO_3 to $30\times500\times500$ nm^3 at 10 wt. % HNO_3). Also, the dealloyed film becomes less dense, because the dealloying rate is proportional to the etchant concentration.

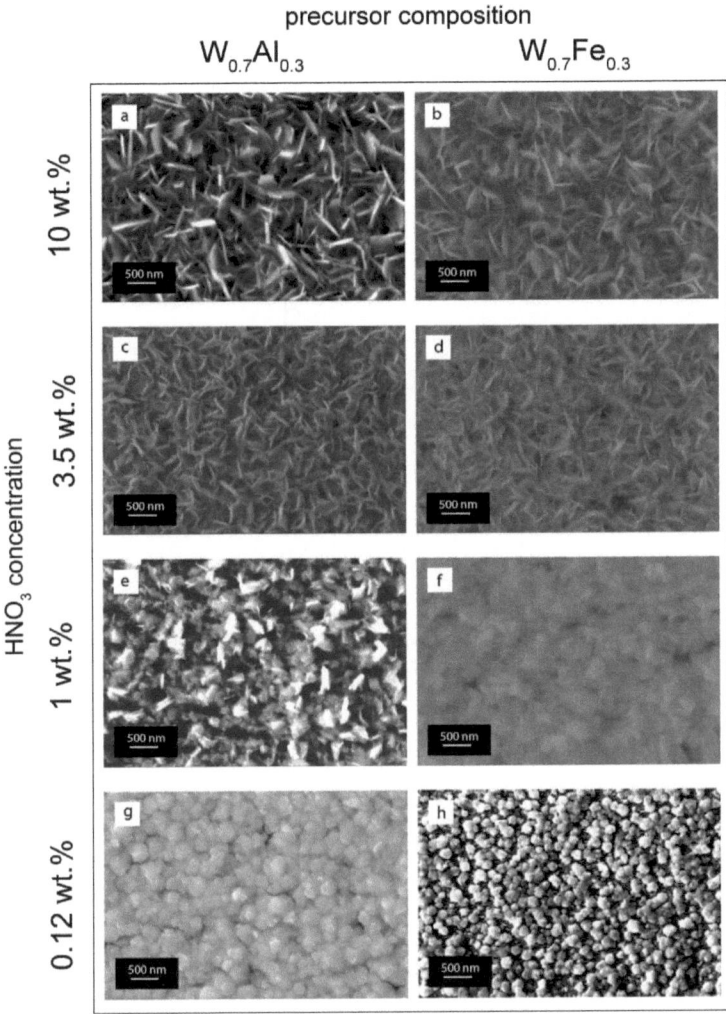

Figure 27: SEM surface topography images of dealloyed $W_{0.7}Al_{0.3}$ (a, c, e, g) and $W_{0.7}Fe_{0.3}$ (b, d, f, h) films. Dealloying of as-deposited films was performed in HNO_3 aqueous solutions with different concentrations: 10 wt.% (a, b), 3.5 wt.% (c, d), 1 wt.% (e, f), and 0.12 wt.% (g, h).

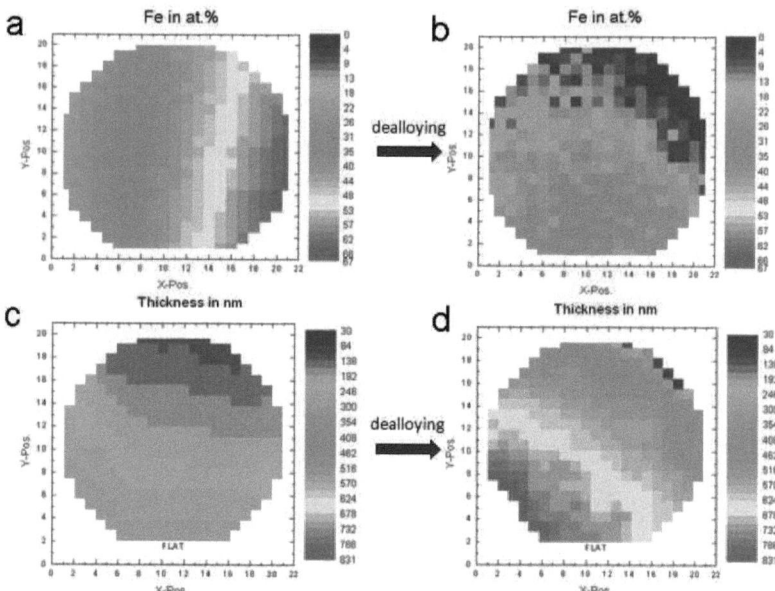

Figure 28: Color-coded composition map (Fe content) of the as-deposited W-Fe materials library before (a) and after (b) dealloying. Film thickness map of the as-deposited W-Fe materials library before (c) and after (d) dealloying.

Moreover, dealloying leads to an increase of the film thickness (on average 3 times; Figure 28c, 28d) due to growth of a highly porous structure, while the content of the soluble component decreases (Figure 28a, 28b). At the same time, an increase of the film thickness is smaller at high concentrations of the soluble component (Al or Fe) in a precursor, as the remaining small amount of W forms thinner films in comparison to W-rich precursors. At low etchant concentrations (c < 0.15 wt. %) the shape of the grains is spherical with diameters of 50-200 nm. At medium concentrations transitional structures are formed. EDX measurements also show that the composition of the films etched at low etchant concentrations (c < 0.15 wt. %) almost does not change.

To explain these phenomena the mechanism of dealloying has to be considered. According to theoretical models [148] the driving force of this process is oversaturation of the interface region between the alloy and the electrolyte with adatoms (here: W adatoms), which appears in the interface region as a result of dissolution of neighboring atoms of the second metal (here either Fe or Al). If the dissolution is performed at high acid concentrations, oversaturation with W adatoms goes faster and a stronger driving force leads to faster W crystal growth. The fast-growing crystals possess high specific surface areas and high surface to volume ratios (Figure 27a, 27b). On the contrary, if diluted acid is used for dissolution, the rate of saturation of the surface layer with W adatoms is small and at most surface areas the oversaturation (required for W grains growth) is not being reached. In this case W adatoms have no driving force for agglomeration and dissolve into the solution faster than oversaturation is reached. Etching of alloys in strongly diluted acid does not lead to dealloying but to simple nonselective etching, leading to exposure of the spherical grains of the precursor film (Figure 27g, 27h).

Figure 29 shows SEM images of $W_{0.7}Fe_{0.3}$, $W_{0.7}Al_{0.3}$ and $W_{0.7}Mg_{0.3}$ thin films etched for 24 h in 0.12 wt. % and 10 wt. % aqueous HNO_3 solutions. Films consisting of cross-linked nanoflakes were fabricated by dealloying the $W_{0.7}Fe_{0.3}$ and $W_{0.7}Al_{0.3}$ films in acid solutions with concentrations of HNO_3 ranging from 3 to 10 wt. %. In case of the $W_{0.7}Mg_{0.3}$ precursor thin film, a formation of nanoflakes is observed even at lower acid concentrations (0.12 wt. %). The growth of flake-like W crystals in the case of W-Mg precursors at much lower acid concentrations than in the case of W-Fe and W-Al precursors indicates that the threshold concentration decreases with decreasing standard electrode potential of the soluble component of a precursor composition.

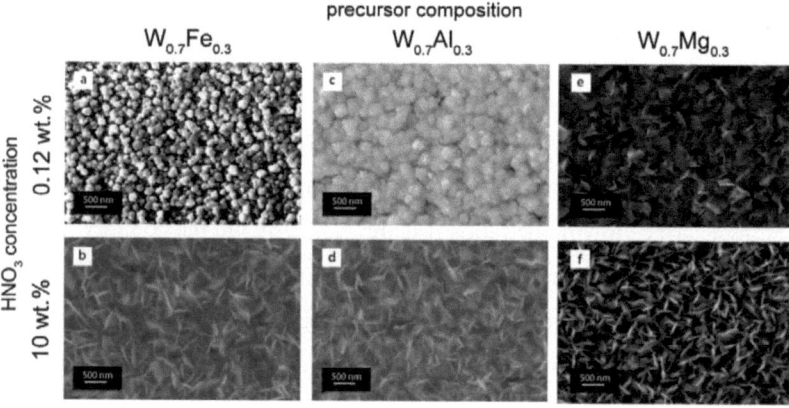

Figure 29: SEM surface topography images of dealloyed $W_{0.7}Fe_{0.3}$ (a, b), $W_{0.7}Al_{0.3}$ (c, d) and $W_{0.7}Mg_{0.3}$ (e, f) films. Dealloying was performed at HNO_3 aqueous solutions with different concentrations: 0.12 wt.% (a, c, e) and 10 wt.% (b, d, f) for 24 h.

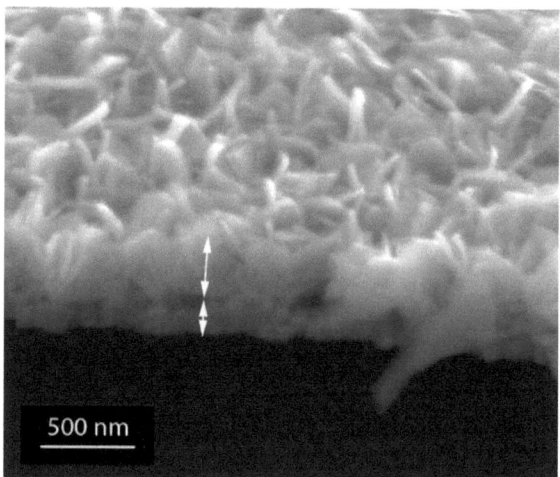

Figure 30: SEM cross-sectional image of the film synthesized by dealloying of the as-deposited $W_{0.7}Fe_{0.3}$ precursor film in 3.5 wt.% HNO_3 aqueous solution for 24 hours. Two layers (dealloyed and not dealloyed) visible at the cross-section are marked with arrows.

The cross-section images (Figure 30) of the dealloyed W-Fe film show its bilayer structure: a porous upper layer consisting of W nanoflakes and an underlying dense layer consisting of the not-dealloyed precursor film. The dense layer plays an important role for the stabilization of the film: when increasing the time of dealloying this layer becomes thinner and when it completely disappears the film is destroyed because the porous dealloyed layer has no adhesion to the substrate.

Comparison of the results of dealloying of W-Al and W-Fe precursor films show that the chemical nature of the second component (Al or Fe) of the precursor alloy does not have any significant impact on the nanostructure of the dealloyed film. The influence of the chemical composition of the film becomes noticeable only after annealing, because at certain compositions annealing leads to the formation of intermetallic compounds. Intermetallic compounds are more stable than solid solutions and the mechanism of dealloying of intermetallic compounds has specific features. Figure 31 depicts the nanostructure of the Fe_7W_6 (intermetallic compound) film after dealloying: it is non-uniform and the nanoflakes are mostly agglomerated to islands. This structure is significantly different from the structure of the dealloyed film with the same composition but not annealed. The nanostructure as observed in SEM images looks similar at different precursor compositions, but at high concentrations of the soluble component (Al or Fe) in the precursor the film becomes more discontinuous, loose and less stable (Figure 32).

Figure 31: SEM surface topography image of the film synthesized by dealloying of intermetallic compound (W_6Fe_7) precursor film (annealed in vacuum at 950°C for 10 hours) in 3.5 wt.% HNO_3 aqueous solution for 24 h.

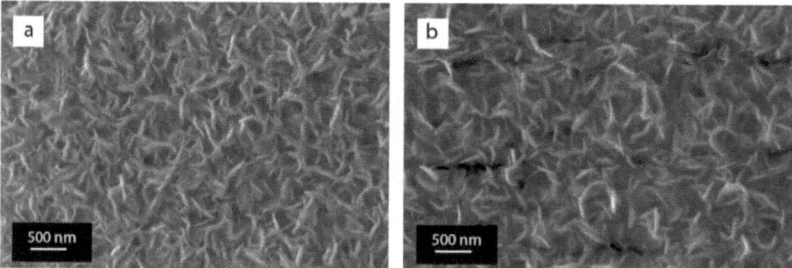

Figure 32: SEM surface topography images of dealloyed $W_{0.8}Al_{0.2}$ (a) and $W_{0.5}Al_{0.5}$ (b) films. Dealloying was performed with 3.5 wt.% HNO_3 aqueous solution for 24 h.

4.1.3 Time-dependent evolution of nanostructure and composition during dealloying

To examine the effects of dealloying and to understand the influence of etching duration on the evolution of nanostructures, as-deposited films were immersed in aqueous HNO_3 (10 wt.%) solution for various times. In an acidic solution, the standard electrode potentials for W^{6+}/W and Fe^{3+}/Fe are 0.680 V and - 0.037 V (standard hydrogen electrode, SHE), respectively. Accordingly, W is thermodynamically more stable than Fe by 0.717 V. As a result, during the dealloying process, the relatively less noble Fe atoms will be selectively dissolved from the W-Fe film.

Figure 33 shows top-view micrographs of the as-deposited film at different compositions. Figures 34 and 35 show that after t_d = 5 min and t_d = 20 min no significant changes in the overall microstructure of the film were distinctly observable. The initiation of the chemically driven dealloying was noticed after t_d = 60 min, where diminutive features evolved at the surface of the W-Fe film. Figure 36 shows top-view micrographs of the film after t_d = 60 min. Evolved nanoflakes on top of the film can be observed. This evidently suggests that the selective dissolution of Fe occurs from the W-Fe precursor film during the dealloying process. For t_d = 180 min., dealloying caused further growth of these structures, as shown in Figure 37. In addition, extensive cross-linking of these structures can be seen. With increasing dealloying duration, selective dissolution of Fe continues and W clusters agglomerate at the solid/electrolyte interface.

Figure 33: SEM surface topography images of as-deposited W-Fe films at different compositions: (a) $W_{0.48}Fe_{0.52}$; (b) $W_{0.62}Fe_{0.38}$; (c) $W_{0.75}Fe_{0.25}$.

Figure 34: SEM surface topography images of $W_{0.48}Fe_{0.52}$ (a), $W_{0.62}Fe_{0.38}$ (b) and $W_{0.75}Fe_{0.25}$ (c) films after dealloying with 10 wt. % HNO_3 aqueous solution for 5 min.

Figure 35: SEM surface topography images of $W_{0.48}Fe_{0.52}$ (a), $W_{0.62}Fe_{0.38}$ (b) and $W_{0.75}Fe_{0.25}$ (c) films after dealloying with 10 wt. % HNO_3 aqueous solution for 20 min.

Figure 36: SEM surface topography images of $W_{0.48}Fe_{0.52}$ (a), $W_{0.62}Fe_{0.38}$ (b) and $W_{0.75}Fe_{0.25}$ (c) films after dealloying with 10 wt. % HNO_3 aqueous solution for 60 min.

Figure 37: SEM surface topography images of $W_{0.48}Fe_{0.52}$ (a), $W_{0.62}Fe_{0.38}$ (b) and $W_{0.75}Fe_{0.25}$ (c) films after dealloying with 10 wt. % HNO_3 aqueous solution for 180 min.

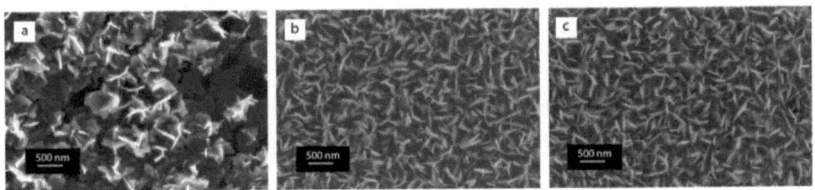

Figure 38: SEM surface topography images of $W_{0.48}Fe_{0.52}$ (a), $W_{0.62}Fe_{0.38}$ (b) and $W_{0.75}Fe_{0.25}$ (c) films after dealloying with 10 wt. % HNO_3 aqueous solution for 1440 min.

Figure 38 shows top-view micrographs of the films dealloyed for t_d = 1440 min. The aggregation of W clusters led to growth and formation of nanoflake structures, on top of the precursor layer. Cluster agglomeration enhanced with respect to dealloying duration and the overall film thickness increased (see Figure 39) due to the growth of a porous, sponge-like structure. In the case of Fe-rich precursors after dealloying for t_d = 1440 min the film becomes very loose and discontinuous (Figure 38a) due to dissolution of a significant portion of the alloy. This leads to delamination and destruction of the film.

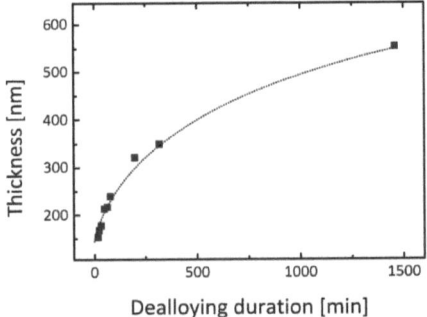

Figure 39: Estimated film thickness of as-deposited and dealloyed films as a function of dealloying duration. The dotted line is a guide to the eyes. Dealloying of 150 nm thick $W_{0.65}Fe_{0.35}$ film was performed with 10 wt. % HNO_3 aqueous solution at room temperature (20°C).

The elemental composition of as-deposited and dealloyed films was studied by EDX. Figure 40 shows color-coded composition maps illustrating W concentration (at. %)

within each W-Fe strip-like materials library with respect to the dealloying duration. With increasing dealloying duration the concentration of Fe drastically reduced, as a result of selective dissolution. The plot in Figure 41 demonstrates elemental atomic composition of the 150 nm thick $W_{0.65}Fe_{0.35}$ film as a function of dealloying duration.

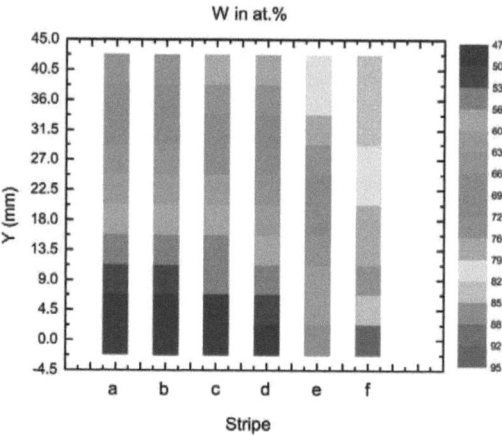

Figure 40: Color-coded composition map (W content) of the W-Fe materials library strips: (a) as-deposited, after dealloying for (b) 5 min., (c) 20 min., (d) 60 min., (e) 180 min., (f) 1440 min. Dealloying was performed at 10 wt. % HNO_3 aqueous solution at room temperature (20°C). The thickness of the as-deposited film was 260 nm.

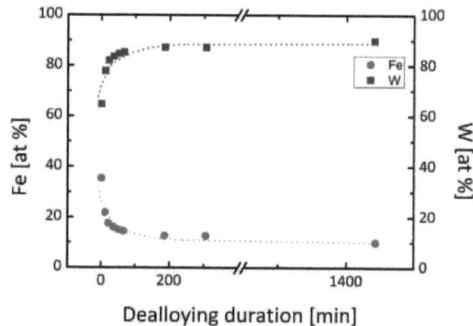

Figure 41: Elemental atomic composition of as-deposited and dealloyed films as a function dealloying duration. The dotted lines are guides to the eyes. Dealloying of the 150 nm thick $W_{0.65}Fe_{0.35}$ film was performed with 10 wt. % HNO_3 aqueous solution at room temperature (20°C).

According to Erlebacher et al. [148], during the dealloying process the dissolution of the less noble metal (here: Fe) occurs, which ultimately leads to the aggregation of the more noble metal (here: W adatoms) into clusters at the solid/electrolyte interface. Throughout this process, the total surface area continuously increases. In the present case, dissolution of neighboring Fe atoms from the W-Fe nanocolumns could cause rapid diffusion of W adatoms towards the interfacial region from the exposed crystalline grains. Subsequently, agglomeration and self-assembly of W adatoms results in the further enhancement of these structures. The nanoflakes exhibit smooth crystalline facets with approximate widths and lengths of about 300-500 nm. Especially, the thickness of the nanoflakes resembles each other to a considerable extent (~ 40 ± 10 nm). A close inspection of the cross-section SEM micrograph suggests growth and formation of smaller flakes, at the base of large nanoflakes. After dealloying the overall film thickness increased 2-4 times, depending on precursor film thickness and composition. In the case of high precursor film thickness only the surface layers of about 150-200 nm were dealloyed during 24 h, but deeper layers remained dense and did not make a contribution to film thickness growth. The highest relative film thickness growth was observed in the case of W-rich thin films, because in this case there was the biggest amount of W needed for the growth of the nanoflakes. Clearly, high growth of the film thickness signifies substantial augmentation of the effective surface area. The chemically driven dealloying increases the specific surface area as a result of dynamic pattern formation and aggregation of the more noble adatoms into two-dimensional clusters at the solid/electrolyte interface [148, 149]. Besides, it is apparent that aggregation of W adatoms and the resultant nanostructures strongly depends on the etching duration.

Although, dissolution of Fe leads to the formation of porous structures [38], diffusion and subsequent rearrangement of W adatoms is a crucial parameter that considerably influences the resulting film nanostructure. According to Erlebacher et al. [149, 150], more noble atoms diffuse out of the base of pits and form clusters to prevent supersaturation of the surface with adatoms, which otherwise could prevent further dealloying. The agglomerates have a tendency of cross-linking, which continuously exposes the alloy to the electrolyte. The present experiments show the thickness decrease of the dense precursor layer and extensive cross-linking of nanoflakes. According to Mullins [151], the adatom diffusion is proportional to the gradient of surface curvature and it occurs in order to minimize the surface energy. Specifically, if agglomerates bear undercuts, the curvature gradient is critically important [150]. In addition, with dissolution of Fe from the W-Fe precursor layer, W adatoms undergo surface self-diffusion over the aggregated clusters. More and more W adatoms accumulate at the solid/electrolyte interface primarily forming small W grains, which further grow into nanoflakes. Therefore, subsequent rearrangement and self-assembly of W adatoms may take place to flatten the surface in order to minimize the surface energy. Particularly, formation of well-defined nanoflakes also indicates a probable direction of adatom diffusion. This adatom diffusion direction could be additionally associated with two simultaneous processes induced by dealloying. First, in order to prevent surface supersaturation with more noble atoms [150], W adatoms may diffuse out of the W-Fe precursor layer to form clusters, in order to continuously expose the alloy. Besides, dissolution of Fe would occur from outer-layer-to-inner-layer, yielding a continuous flux of W adatoms. Secondly, the dissolved Fe atoms undergo a directional flow outward towards the electrolyte. Therefore, it is plausible that the diffusing W adatoms experience a strong thrust for subsequent aggregation and self-assembly. Further, the aggregates flatten to minimize the surface energy [151], which could lead to the growth of large nanoflakes.

The crystallographic properties of as-deposited and dealloyed W-Fe films were determined by XRD measurements. A reference as-deposited W-Fe precursor film was examined for comparison. Figure 42 shows XRD patterns of as-deposited and dealloyed films with corresponding compositions. For as-deposited and dealloyed W-Fe films high intensity β-W(200) and α-W(110) reflections are observed. No reflections that can be ascribed to Fe were noticed for the as-deposited and dealloyed W-Fe films. For the dealloyed W-Fe films (t_d = 5 to 1440 min), WO_3 (200) and (220) peaks are observed, signifying the presence of the monoclinic WO_3 phase. Moreover, with increasing dealloying duration the intensity the WO_3 peaks increased. This indicates partial oxidation of the film during the dealloying process.

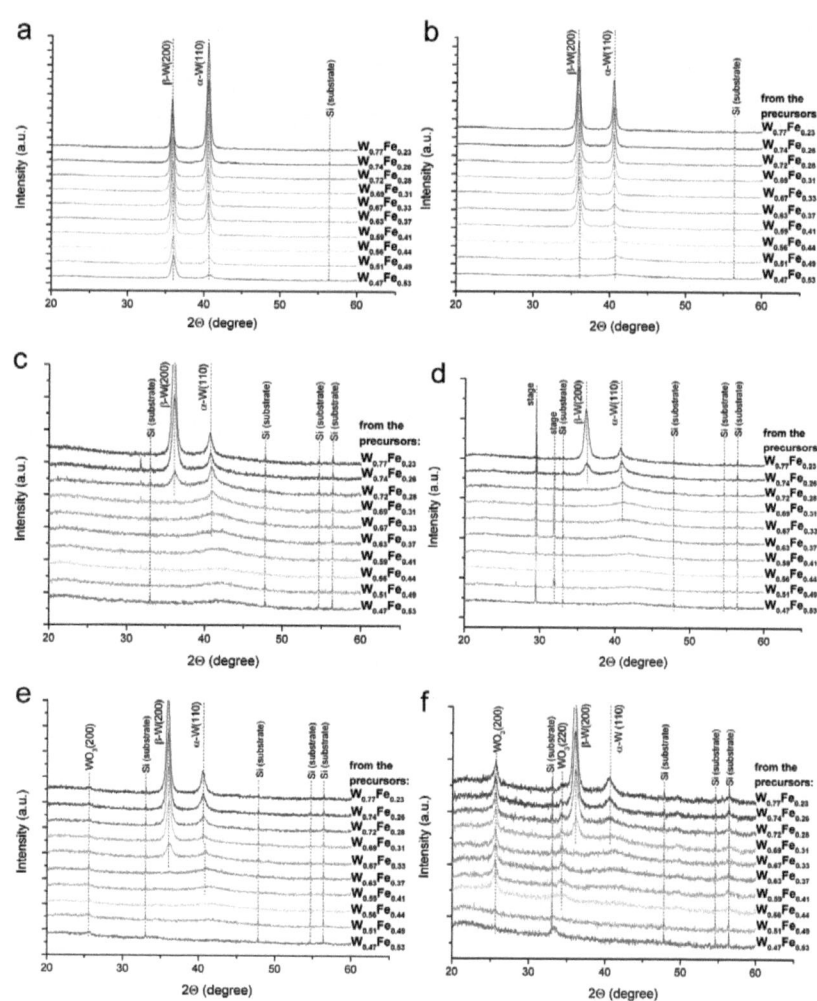

Figure 42: XRD patterns of W-Fe films: as-deposited (a) and dealloyed for 5 min. (b), 20 min. (c), 60 min. (d), 180 min. (e) and 1440 min. (f) with 10 wt. % HNO_3 aqueous solution at room temperature (20°C).

To further assess the local crystallographic structure of dealloyed W-Fe films, TEM investigations were performed. For local crystallographic analysis of nanoflakes, a small part of the sample was scraped and transferred onto a TEM grid. Figure 43 shows an example of one such rectangular nanoflake that formed after the dealloying process (t_d =1440 min.). The selected area electron diffraction (SAED) pattern (shown in inset) evidently reveals that the single crystal nanoflake is comprised of the monoclinic WO_3 phase. The single crystal nanoflake was found evolve along the [002] and [020] axes, with longer growth in [002] direction. The length and width of the nanoflake measured ~ 370 nm and ~ 257 nm, respectively. The present TEM observations are consistent with the earlier discussed SEM and XRD investigations. Thus, the dealloying demonstrates a simple route to fabricate single crystal WO_3 nanoflakes.

Figure 43: Bright field TEM micrograph[10] of a single crystal WO_3 nanoblade formed by dealloying a W-Fe nanocolumnar film for t_d = 1440 min. The inset shows the selective area electron diffraction (SAED) pattern obtained from the nanoblade, indicating the monoclinic phase of the WO_3 single crystal.

---

[10] TEM image by Dr. Pio John Buenconsejo

4.1.4 Oxidation of the dealloyed films

The dealloyed films were further annealed at 500 °C in air for 8 h with heating and cooling rates of 2 K/min., to complete the oxidation of the remaining metallic W into WO_3. Figure 44 shows a SEM image of the dealloyed $W_{0.7}Fe_{0.3}$ films before (a) and after oxidation (b). This demonstrates that after annealing the porous structure remains intact, with only the nanoflakes becoming slightly thicker. The XRD patterns shown in Figure 45 suggest that all oxidized samples consist of polycrystalline monoclinic WO_3. At the same time, film thickness does not change during the oxidation process because there is enough space in between nanoflakes for volume expansion due to the high porosity. Also one should take into account that the films after dealloying in HNO_3 are already partially oxidized (see Chapter 4.1.6). Thus, further annealing in air is needed just to complete the oxidation.

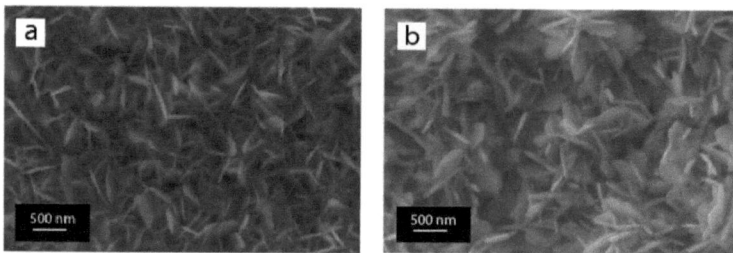

Figure 44: SEM images of the porous films synthesized by dealloying of $W_{0.7}Fe_{0.3}$ precursor films with 10 wt. % HNO_3 aqueous solution for 24 h: (a) before and (b) after thermal oxidation at 500 °C in air for 8 h.

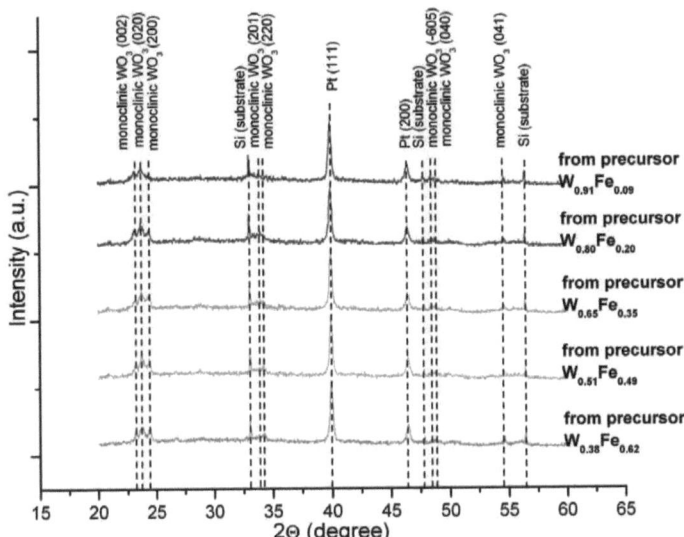

Figure 45: XRD patterns of WO$_3$ porous films synthesized by thermal oxidation of dealloyed W$_x$Fe$_{1-x}$ precursor films with different W:Fe ratios.

4.1.5 Photoelectrochemical properties of the dealloyed W-Fe dense films

Photocurrent was used as the screening parameter, for characterization of PEC activity of the electrodes. This approach and its advantages were described in detail in chapter 3.2.6. Photocurrent measurements were performed by the PEC scanning droplet cell, as described in chapter 3.2.6. Figure 46 shows the steady-state photocurrent density map of a complete materials library measured at fixed illumination intensity.

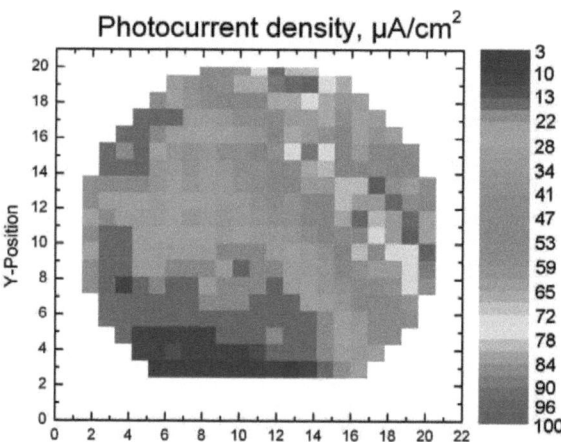

Figure 46: Color-coded photocurrent density map of the porous WO₃ materials library fabricated by dealloying of a W-Fe precursor materials library with composition and thickness gradients (see Figure 28).

Highly porous films showed significantly higher (up to 40 times) photocurrent responses compared to dense films. The recombination of electron-hole pairs is sufficiently suppressed in porous films. Importantly, in contrast to dense electrodes, the photocurrent generation in porous semiconductor electrodes is not simply limited by the diffusion length of the charge carriers (hole diffusion length in WO₃ ~150 nm, [21]) since the pores are filled with electrolyte and even the holes generated deep in the film can react with water. In other words, the porous structure enhances the photoconversion efficiency by rendering the path of photogenerated holes into the solution shorter and thus avoiding recombination. Film thickness is one more crucial parameter. In the case of thick films the path of charge carriers is too long and leads to large losses due to the recombination of electron-hole pairs. On the other hand, too thin photoelectrodes produce low photocurrent densities, because light absorption efficiency decreases with decreasing film thickness. These two mechanisms explain the existence of an optimal film thickness, for which the balance between light absorption and charge carriers recombination is reached.

To understand which factors influence the PEC properties of the dealloyed films, one has to investigate in more detail nanostructure and composition, taking into account the change of porosity and composition along the film thickness, and consider three dependencies: (1) how relative film thickness growth during dealloying depends on the precursor film thickness and precursor composition, (2) how the composition of dealloyed films depends on the precursor composition and thickness, (3) how photocurrent depends on the content of impurities (Fe, Al, Mg) in the dealloyed film. The electrodes produced from W-Fe, W-Al and W-Mg precursors demonstrated similar values of photocurrent (maximum up to 100 $\mu A/cm^2$, 104 $\mu A/cm^2$ and 106 $\mu A/cm^2$ respectively) but possess different mechanical and corrosion stability. Mechanical stability was evaluated by adhesion of the film to the substrate, presence/absence of macro cracks and flaking during or after PEC measurements. Corrosion stability was determined by evaluating the current density during amperometric measurements without illumination. The stability of the electrodes was the best in the case of W-Fe and the worst in the case of W-Mg, which clearly correlate with the standard electrode potential of the soluble component. On the one hand, in the case of a chemically very active second component in the precursor, dealloying is faster and the growing porous nanostructure accumulates more defects and therefore becomes less stable. On the other hand the active elements remaining in the film after dealloying, dissolve easier during photocurrent measurements, which leads to electrode degradation and finally destruction. Taking into account all of these factors, the W-Fe precursor system was chosen for further investigations.

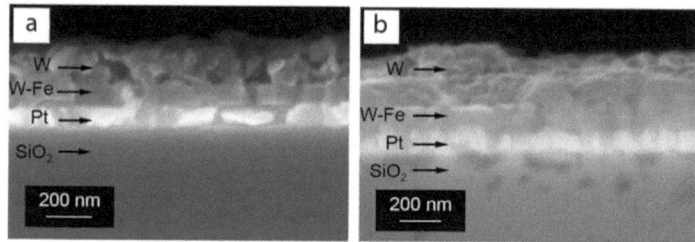

Figure 47: SEM cross-sectional images of the film synthesized by dealloying of the as-deposited 130 nm thick $W_{0.6}Fe_{0.4}$ precursor film (a) and 200 nm thick $W_{0.8}Fe_{0.2}$ precursor film (b) in 10 wt. % HNO_3 aqueous solution for 24 h with further annealing at 500 °C in air for 8 h.

The cross-section images (Figure 47) of the dealloyed W-Fe film show its bilayer structure in dependence on precursor film thickness and composition. Two layers (porous dealloyed, and dense not-dealloyed) of the precursor, Pt layer (deposited in order to provide high conductivity) and the SiO_2 barrier layer on top of the Si substrate are visible in the cross-section. Firstly, this shows that the porosity of dealloyed films is higher when the precursor contains lower amounts of W, resulting in formation of a smaller amount of better separated nanoflakes. Secondly, this shows that the thickness of the dense layer is proportional to the initial precursor film thickness.

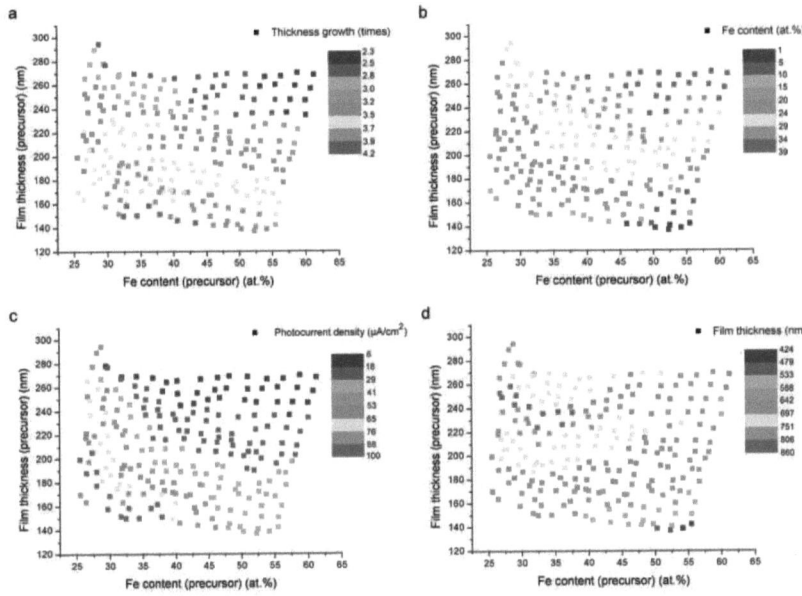

Figure 48: Color-coded diagrams of relative film thickness growth (a), Fe content after dealloying (b), photocurrent density (c), film thickness after dealloying (d) in dependence on precursor film thickness and composition.

The diagrams in Figure 48 were made to answer the question about the influence of the dense layer on PEC properties. Figure 48a shows that higher relative film thickness growth is observed in the case of smaller precursor film thickness. This means that thicker films are not completely dealloyed and always contain a dense layer underneath. The thickness of the dense layer is proportional to the thickness of the precursor film. Relative film thickness growth also reflects porosity and could be used as a parameter for porosity comparison. Figure 48b shows that the remaining Fe content in the dealloyed film is also proportional to the precursor film thickness. This means that the Fe content depends on the thickness of the dense layer and Fe is located almost exclusively in the dense layer. Taking into account these conclusions and Figure 48c, a strong correlation of photocurrent density with film porosity, the thickness of the dense layer and the Fe content in it was observed. The highest photocurrent densities were observed at small thicknesses of the dense layer (0 - 50 nm) and at relatively small Fe content in it (< 38 at.%), while the overall Fe content in the film (including dense and porous layers) must be even lower (< 18 at.%). At the same time, the photocurrent density (Figure 48c) does not exhibit a good correlation with the overall film thickness (Figure 48d). This means that the dense layer thickness and its composition are critical parameters. With increasing dense layer thickness, electron-hole pairs recombination increases and a large portion of charge carriers recombine before they reach the conductive metallic layer of the electrode. High Fe content in the dense layer also decreases photocurrent density, because Fe_2O_3 has worse charge transfer properties than WO_3 [32, 152]. Therefore, theoretically the highest photocurrent densities would be observed without a dense layer. However, film destruction was observed after longer etching when the dense layer fully disappears. This means that the dense layer plays the role of an adhesion layer between the porous film and the substrate and provides the necessary mechanical stability of the film.

4.2 Ti-based systems: dense film precursors

Figure 49 shows a series of composition-dependent XRD patterns of the as-deposited Ti-Mg thin film materials library. Peaks corresponding to α-Ti are observed for all compositions. Peaks corresponding to α-Mg and metastable β-Ti phases appear at Mg-rich compositions.

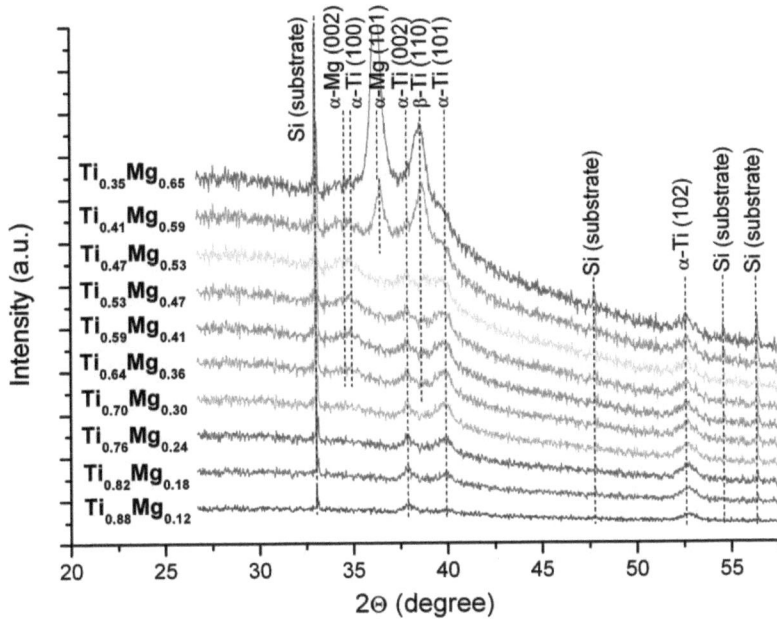

Figure 49: XRD patterns from the as-deposited Ti-Mg materials library for measurement regions with different Ti:Mg ratios.

Figure 50 shows the surface topography of dealloyed Ti-Mg thin films. At different compositions (from $Ti_{0.7}Mg_{0.3}$ to $Ti_{0.5}Mg_{0.5}$) and different etchant concentrations, formation of spherical Ti grains (40-80 nm in diameter) is observed due to selective Mg dissolution. The structure is denser in comparison to dealloyed W-based precursors, because spherical Ti grains grow compactly to each other as compared to lamellar W grains which form a very loose sponge-like structure. In contrast to W-based precursors, the thickness of Ti-based films does not increase after dealloying, due to compact positioning of the spherical grains. At compositions with high Mg content, the film becomes discontinuous consisting of interconnected islands due to dissolution of a large portion of the material.

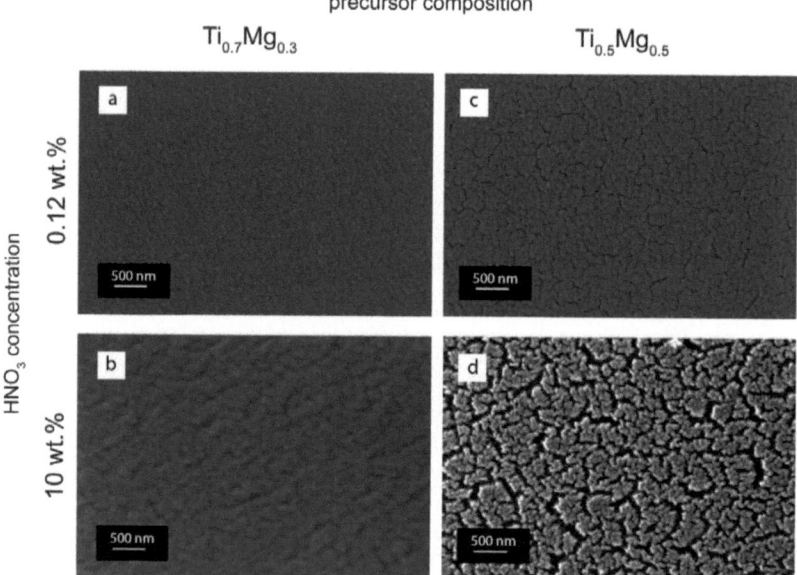

Figure 50: SEM surface topography images of dealloyed $Ti_{0.7}Mg_{0.3}$ (a, b) and $Ti_{0.5}Mg_{0.5}$ (c, d) films. Dealloying was performed with HNO_3 aqueous solutions with different concentrations: 0.12 wt. % (a, c) and 10 wt. % (b, d) for 24 h.

After dealloying and thermal oxidation at 500°C, the XRD patterns show peaks corresponding to rutile (α-TiO_2) (Figure 51).

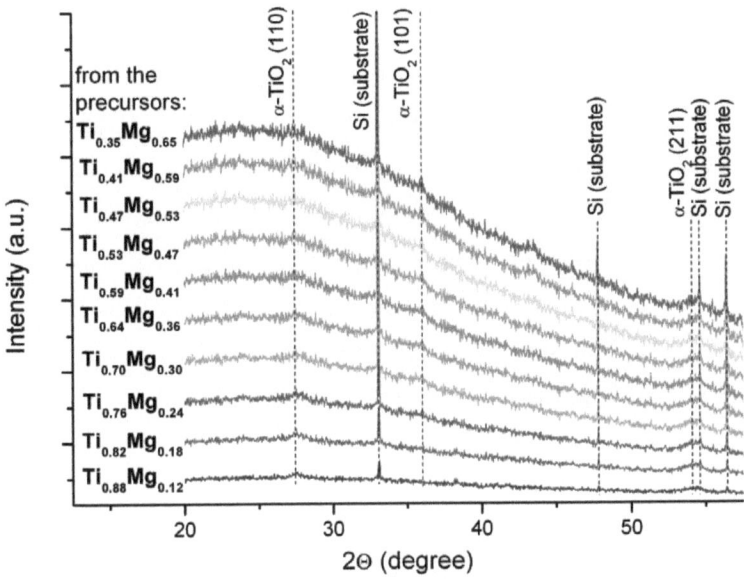

Figure 51: XRD patterns of a TiO$_2$ film synthesized by thermal oxidation (500°C, 8 h) of dealloyed Ti$_x$Mg$_{1-x}$ precursor films with different Ti:Mg ratios.

During amperometrical measurements at 1 V applied bias very high dark current values (up to 250 µA/cm^2) were observed at Mg-rich TiO$_2$ photoanodes, which is an evidence of corrosion. For comparison, in the case of WO$_3$ anodes with low content of the soluble components (Mg, Al, or Fe), dark current values were close to zero. Thus, the presence of high amounts of Mg in photoanodes was shown to be undesirable, as it decreases the mechanical and corrosion stability of electrodes.

4.3 W-Ti-Mg ternary system: dense film precursors

Figure 52 shows a series of composition-dependent XRD patterns of the as-deposited ternary W-Ti-Mg thin film materials library. At all compositions only the peaks corresponding to α-W are observed. The formation of the W-based solid solution becomes possible in a broad range of compositions. As known from literature, W often forms solid solutions with the crystalline structure of bcc W by substituting W atoms with other elements (e.g. Ti) [145]. Since the atomic radii of Ti and Mg are higher than that of W, the substitution of W by Ti and Mg leads to dilatation of the lattice, which is clearly visible by α-W(110) peak shift when increasing Ti and Mg content.

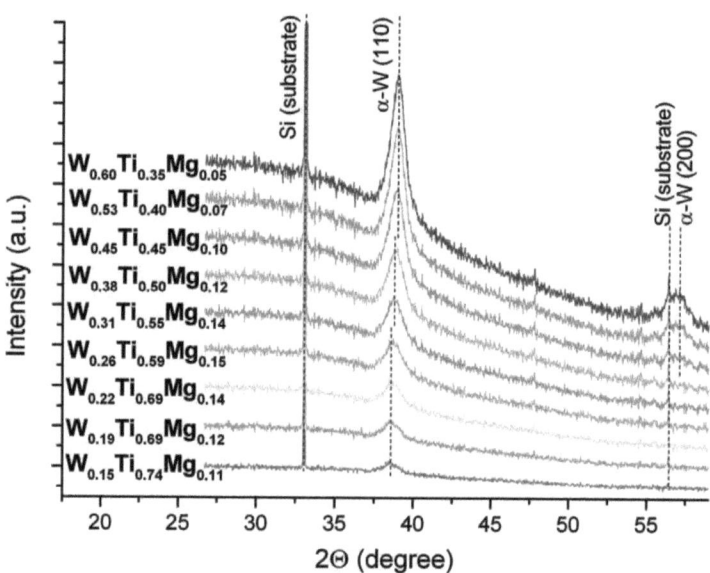

Figure 52: XRD patterns from the as-deposited W-Ti-Mg materials library for measurement regions with different W:Ti:Mg ratios.

A phase separation effect was observed during dealloying of W-Ti-Mg precursors with high W content (Figure 53). At high W concentrations in the precursor, dealloying leads to formation of a porous W layer on top of a Ti-rich layer due to phase separation, i.e. W crystals grow separately from Ti. At low W content (< 15 at. %) only a uniform Ti-rich layer is formed. At moderate W concentrations (20-30 at. %) separate islands of W consisting of agglomerated flakes appear on top of the Ti-rich layer. For W concentrations >50 at. % the number of W grains increases and they form an almost continuous layer on top of the Ti-rich layer.

Figure 53: SEM surface topography images of dealloyed $W_{0.1}Ti_{0.7}Mg_{0.2}$ (a), $W_{0.25}Ti_{0.55}Mg_{0.2}$ (b) and $W_{0.5}Ti_{0.3}Mg_{0.2}$ (c) films. Dealloying was performed with 10 wt.% HNO_3 aqueous solution for 24 h.

In contrast to W-based alloys, Ti-based alloys are not as good precursors for dealloying. Dealloying of Ti-containing alloys progresses very slowly and less efficiently. At higher Ti contents in a precursor (Ti-Mg or W-Ti-Mg) a higher portion of the soluble metal (Mg) remains not-leached (Figure 54). This phenomenon could be explained by passivation and the low mobility of Ti atoms. Ti is known as strong passivation promoter which suppresses the dealloying processes [153].

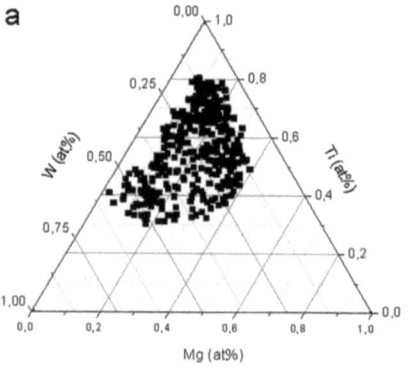

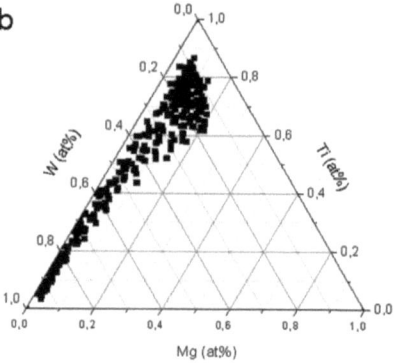

Figure 54: Elemental composition diagram for the system W-Ti-Mg before (a) and after (b) dealloying. Dealloying was performed with 10 wt. % HNO$_3$ aqueous solution for 24 h. Elemental composition was measured by energy dispersive X-ray analysis (EDX).

4.4 W-Fe system: nanocolumnar film precursors
4.4.1 Nanostructure and composition of the nanocolumnar films

As concluded in chapter 4.1.4, the photocurrent density increases with increasing film porosity and decreasing dense layer thickness. Therefore, theoretically the highest photocurrent densities could be observed without any dense layer. However the dense layer provides mechanical stability of the film and so may not be completely avoided. To overcome this limitation, one should increase porosity of the dense stabilization layer. Porosity of the stabilization layer can be increased by using dealloying precursors with higher specific surface. For example one can use nanocolumnar arrays, fabricated by GLAD deposition. In this case the overall porosity of the film after dealloying will be higher due to a double effect: the porosity evolved during the dealloying process and the inherent porosity of the precursor.

As described in chapter 3.1.1, the GLAD geometry of the sputtering chamber includes 2 cathodes positioned at different inclination angles: W oblique angle source and Fe confocal incidence source. First a ~ 100 nm nucleation layer of W nanocolumns was deposited followed by subsequent co-deposition of W and Fe. Thus the deposited film has a bi-layer columnar structure comprising of W and W-Fe layers. Figure 55a and 55b show cross-section and top-view SEM micrographs respectively of as-deposited W/W-Fe nanocolumns. Since co-deposition was carried out with an oblique angle source (W: $\beta \approx 88°$) and a confocal incidence source (Fe: $\beta \approx 27°$) well-separated, columnar nanostructures evolved due to the self-shadowing effect.

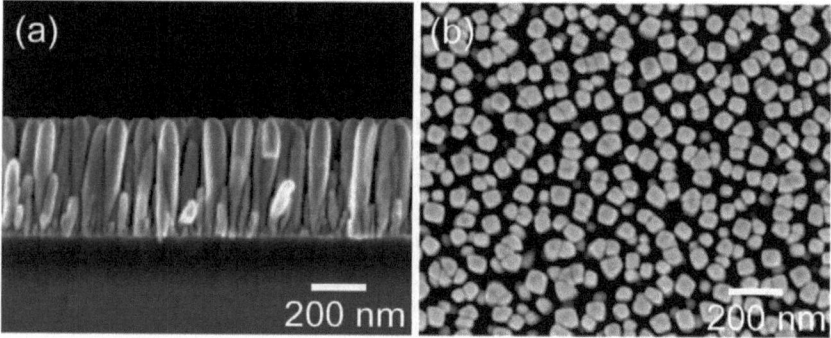

Figure 55: Cross-sectional (a) and top-view (b) SEM micrographs of as-deposited $W_{65}Fe_{35}$ nanocolumns films deposited using the glancing angle co-deposition method with an oblique angle source and confocal incidence source.

As the W-Fe co-deposition process continues on the W nanocolumnar nucleation layer (h ~ 100 nm), the predominant columns receive more particle flux as a result of a competitive growth mechanism [154]. Under negligible surface diffusion conditions, the impinging particles will stick where they land on the surface of the W nucleation layer. The inherent shadowing mechanism during the GLAD process therefore leads to the formation of vertical columnar entities, perpendicular to the substrate surface. The $W_{0.65}Fe_{0.35}$ composition was used for fabrication of nanocolumnar precursors, because it showed the best performance in the case of dense films (chapter 4.1). The as-deposited $W_{65}Fe_{35}$ film exhibited a total film thickness of 453 nm, an average column width of 77 ± 15 nm, and a column number density of 119 ± 14 μm^{-2} (see Figure 55).

4.4.2 Formation of nanostructures during dealloying

The dealloying was performed in aqueous HNO_3 (10 wt. %) solution for various times, in such a similar way to how was done in the case of dense film precursors (chapter 4.1.3).

The initiation of the chemical dealloying was noticed after dealloying for 15 min., with fine features evolved at the surface of W-Fe nanocolumns (Figure 56), that is much faster than in case of dense film precursors, where the initiation of the chemical dealloying was noticed only after 60 min (see chapter 4.1.3). This demonstrates that the dealloying rate is proportional to the specific surface of the precursor. In the case of columnar films, a larger surface is exposed to the etchant solution, which leads to an increase of the reaction rate.

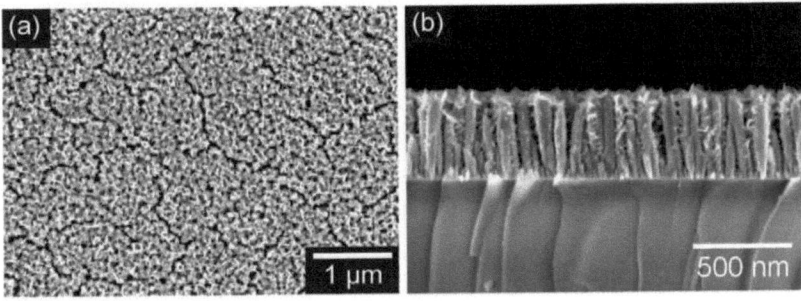

Figure 56: Top-view and corresponding cross-section SEM micrographs of $W_{65}Fe_{35}$ nanocolumnar film after dealloying for 15 min. in aqueous HNO_3 (10 wt. %) solution.

Figure 57a and 57b show top-view and cross-section micrographs of a nanocolumnar film after t_d = 30 min. From Figure 57b nanocacti-like structures evidencing fine features can be observed. It is important to note that the fine features did not evolve from the W columnar nucleation layer. For t_d = 60 min, dealloying caused further coarsening of these structures, as shown in Figure 57c and 57d. After t_d = 300 min., significant alteration in the overall morphology can be observed, with large nanoflakes evolved on top of the nanocolumnar film (see Figure 57e and 57f).

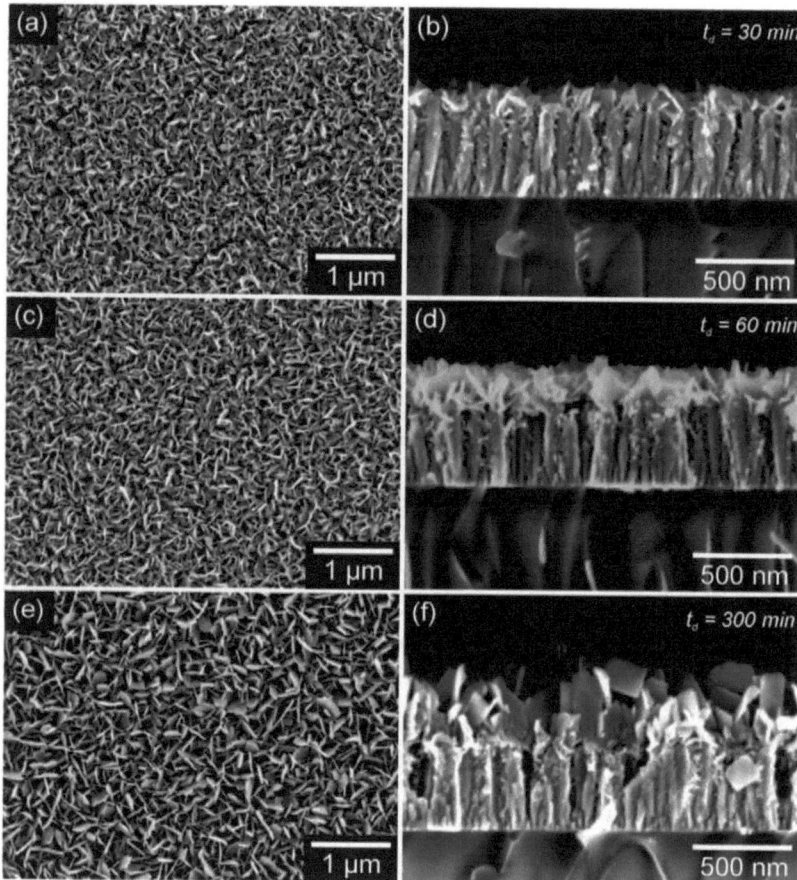

Figure 57: Top-view and corresponding cross-section SEM micrographs showing the evolution of various nanostructures from $W_{65}Fe_{35}$ nanocolumns after dealloying the film in an aqueous HNO_3 (10 wt. %) solution for different time durations: (a-b): t_d = 30 min., (c-d): t_d = 60 min. and (e-f) t_d = 300 min.

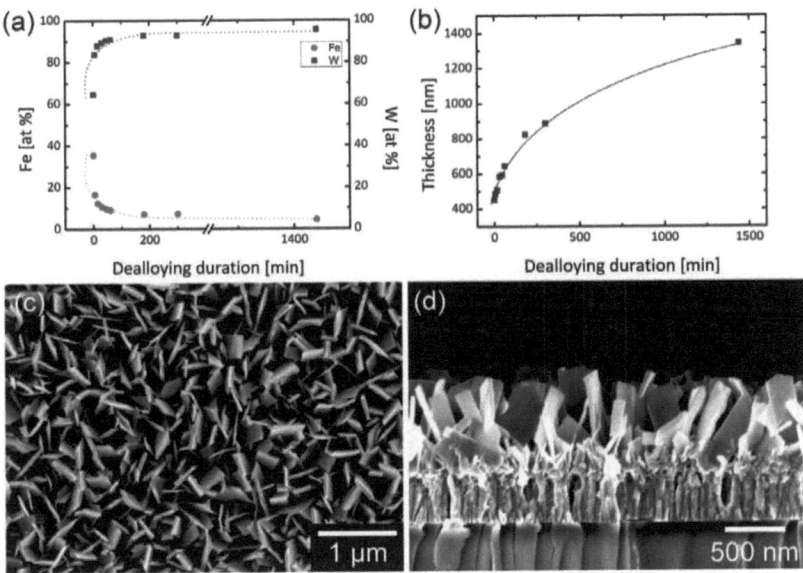

Figure 58: (a) Elemental atomic composition of the as-deposited and dealloyed films as a function of dealloying duration, (b) Estimated film thickness of the as-deposited and dealloyed (aqueous 10 wt.% HNO_3) $W_{65}Fe_{35}$ films as a function of dealloying duration. The dotted lines are guides to the eyes. (c-d) Top-view and corresponding cross-section SEM micrographs showing formation of nanoblades after dealloying $W_{65}Fe_{35}$ films in an aqueous HNO_3 (10 wt. %) solution for $t_d = 1440$ min.

The elemental composition of as-deposited and dealloyed films was studied using EDX. Figure 58a shows a plot of Fe and W concentration (at. %) within each sample with respect to the dealloying duration. With increasing dealloying duration the concentration of Fe was drastically reduced as a result of selective dissolution, from the as-deposited composition $W_{65}Fe_{35}$ to $W_{96}Fe_4$ after $t_d = 1440$ min. In addition, cluster agglomeration enhanced with respect to dealloying duration and the overall film thickness increased to $h \approx 1345$ nm for $t_d = 1440$ min (see Figure 58b). Figure 58c and 58d show top-view and cross-section micrographs of the film after it was dealloyed for $t_d = 1440$ min. The aggregation of W clusters led to growth and formation of nanoblade-like structures, on top of the W columnar nucleation layer.

4.4.3 Crystallographic characteristics of W-Fe nanocolumnar films

The crystallographic properties of as-deposited and dealloyed W-Fe nanocolumnar films were determined by XRD measurements. A reference GLAD W nanocolumnar film (thickness ~ 450 nm) was examined for comparison. Figure 59 shows XRD patterns of as-deposited and dealloyed nanocolumnar films with corresponding compositions. For as-deposited, dealloyed W-Fe films and a reference W nanocolumnar film, very intense β-W(200) and β-W(210) reflections are observed. No reflections ascribed to Fe were noticed for as-deposited and dealloyed W-Fe films. For the dealloyed W-Fe films (t_d = 5 to 1440 min.), WO_3 (020), (200) and (400) peaks are observed, signifying the presence of the monoclinic WO_3 phase, as was the case for dealloying of dense film precursors. With increasing dealloying duration the intensity of the WO_3 peaks increased. This indicates partial oxidation of the film during the dealloying process.

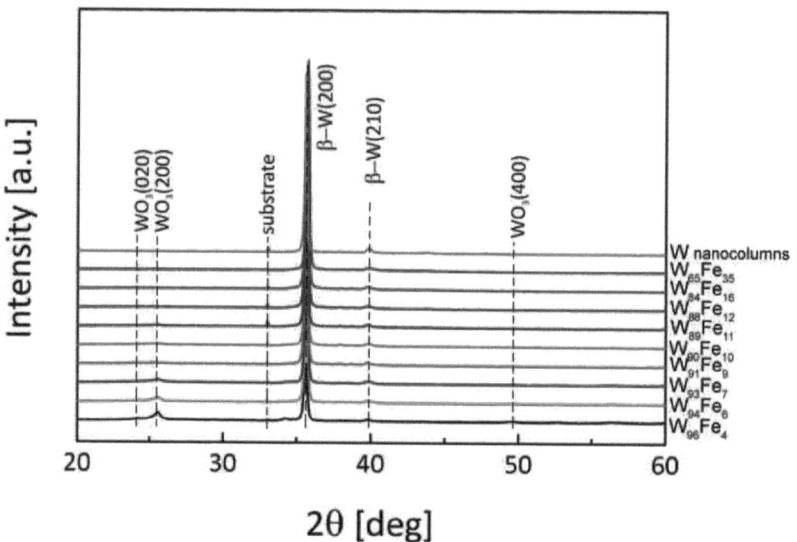

Figure 59: XRD patterns of reference W nanocolumns, as well as as-deposited and dealloyed W-Fe films. For each sample the overall W_xFe_{100-x} composition for the pattern is indicated.

In contrast to dense as-deposited W-Fe films containing both α-W and β-W phases, nanocolumnar films fabricated under GLAD conditions contain only the β-W phase. The explanation of this phenomenon was found by considering the growth mechanisms of the nanocolumnar films. The growth mechanism of α-W and β-W phases has been the question of several studies in the past [155-157]. It is known that α-W is the thermodynamically stable state with higher adatom mobility and a bcc structure, while the metastable β-W phase is considered to have poor adatom mobility [158]. Monte Carlo simulations [159] suggest that obliquely incident particles have an increased probability of getting trapped on islands leading to lower surface mobility. Particularly, under GLAD conditions, islands formed at the initial stages govern the subsequent nanostructure growth. As deposition continues, due to the inherent shadowing mechanism, the predominant islands become taller and shadow minute surface features. Therefore, under such conditions lower mobility β-W islands may grow in comparison to α-W phase. The present observations are also in agreement with the detailed investigation of GLAD-grown β-W phase nanocolumns by Karabacak et al. [160, 161]. However, it is interesting to note that even for co-deposited W-Fe nanocolumnar films the β-W phase is observed in the as-deposited W-Fe film.

4.4.4 Photoelectrochemical properties of the dealloyed W-Fe nanocolumnar films

Before PEC characterization the dealloyed columnar films were oxidized by annealing at 500 °C in air for 8 h, with heating and cooling rates of 2 K/min.

Photocurrent measurements were performed by the PEC scanning droplet cell, as described in chapter 3.2.6. Photocurrent densities measured at dealloyed and oxidized nanocolumnar thin films reach a maximum of 230 $\mu A/cm^2$, which is 2.3 times higher than the maximum photocurrent densities measured for the films synthesized from dense precursors (chapter 4.1.8). At the same time, mechanical stability of the dealloyed columnar films is higher than in the case of the films synthesized from dense precursors, resulting in an important advantage. The current density measured without illumination was close to zero, which is evidence of good corrosion stability of the electrodes. Thus, functional properties of the WO_3 photoelectrodes fabricated from the dealloyed nanocolumnar films is significantly better in comparison to those of dealloyed dense films.

5 Conclusions and outlook

In this thesis, the combinatorial materials science approach was applied for the investigation of dealloying of binary and ternary W- and Ti-based alloys and for the development of porous photoelectrodes for solar water splitting applications. Dense and nanocolumnar thin film materials libraries with continuous composition and thickness gradients were fabricated using PVD. Combinatorial chemical dealloying of the metallic thin film precursors was implemented for fabrication of highly porous photoelectrodes. Materials libraries were investigated by high-throughput characterization techniques: automated EDX analysis for elemental composition characterization, XRD for phase structural characterization, thickness and PEC measurements. Additionally, SEM and TEM were used for nanostructure characterization.

Porous films were produced by dealloying of the binary precursor alloys W-Fe, W-Al, W-Mg, Ti-Mg and the ternary precursor alloy W-Ti-Mg. Evolution of nanostructure and composition during dealloying was investigated in dependence on precursor composition, precursor film thickness and nanostructure, etchant concentration and dealloying duration. In the case of W-based precursors it was found that the etchant concentration must be above a certain threshold, which depends on the standard electrode potential of the components in the precursor. The change of the soluble component of the precursor (Fe, Al, Mg) does not influence significantly the resulting nanostructure and PEC characteristics of the dealloyed and oxidized films, but does influence their mechanical and chemical stability. In the case of soluble components with lower standard electrode potentials in the precursor, lower etchant (HNO_3) concentrations are required to perform dealloying. The highest mechanical and chemical stability of the dealloyed films was found for the case of the least active soluble component (Fe) in the precursor. The dealloying rate was found to be proportional to the etchant concentration. The same resulting effect on nanostructure evolution was observed either by increasing the dealloying duration or increasing the etchant concentration. The rate and yield of dealloying depends on concentration of passivation promoters (such as Ti) in the precursor. Therefore, dealloying of W-Fe, W-Al and W-Mg precursors progresses much faster and with a higher yield than Ti-Mg or Ti-rich W-Ti-Mg. Dealloying of W-Ti-Mg with high W concentrations leads to the formation of a porous W layer on top of a Ti-rich layer due to phase separation.

The dealloyed films have a bilayer structure with a porous upper layer and a dense not-dealloyed layer underneath. Photocurrent densities strongly depend not only on general film porosity, but also on dense layer thickness and composition. The dense layer thickness has much stronger influence on the photocurrent than does the overall film thickness. The photocurrent density increases when decreasing the dense layer thickness and the Fe, Al or Mg content in it. Therefore, theoretically the highest photocurrent densities would be observed without any dense layer. However the dense layer provides mechanical stability of the film and may not be completely avoided. This problem was solved by using nanocolumnar thin film precursors for dealloying instead of dense thin film precursors. That allows elimination of the dense layer, increases the overall film porosity and improves mechanical stability of the dealloyed films. It helped to increase photocurrent densities up to 2.3 times in comparison to the maximum photocurrent densities measured in the films synthesized from the dense precursors.

The present results show the advantages and convenience of dealloying for fabrication of porous thin film materials, important for photoelectrochemical applications. The extension of this approach to other multi-element systems in combination with combinatorial and high-throughput techniques can be envisioned. In order to minimize the costs of precursor fabrication, cheaper electrochemical techniques for film depositions should be used.

6 References

1. *"Shell Energy Scenarios to 2050: Signals & Signposts"* Shell International BV 2011 [cited 2013 14th February]; Available from: www.shell.com/scenarios.
2. Armaroli, N. and V. Balzani, *The future of energy supply: Challenges and opportunities.* Angewandte Chemie-International Edition, 2007. **46**(1-2): p. 52-66.
3. Armaroli, N. and V. Balzani, *Energy for a Sustainable World: From the Oil Age to a Sun-Powered Future* 2010, Weinheim: Wiley-VCH.
4. Olah, G.O., A. Goepert, and G.K.S. Prakash, *Beyond Oil and Gas: The Methanol Economy* 2009, Weinheim: Wiley-VCH.
5. Holdren, J.P., *Global environmental issues related to energy supply - the environmental case for incresed efficiency of energy use.* Energy, 1987. **12**(10-11): p. 975-992.
6. Armaroli, N. and V. Balzani, *The Hydrogen Issue.* Chemsuschem, 2011. **4**(1): p. 21-36.
7. Oxtoby, D.W., *Principles of Modern Chemistry (5th ed.)* 2002: Thomson Brooks/Cole.
8. *"Hydrogen Properties, Uses, Applications"* Universal Industrial Gases, Inc. . 2007 [cited 2013 14th February]; Available from: http://www.uigi.com/hydrogen.html.
9. Ludwig, A., et al. *Development of new materials for hydrogen production and storage using high-throughput methods.* in *1st International Conference on Materials for Energy.* 2010. Karlsruhe.
10. Balzani, V., A. Credi, and M. Venturi, *Photochemical conversion of solar energy.* Chemsuschem, 2008. **1**(1-2): p. 26-58.
11. Osterloh, F.E. and B.A. Parkinson, *Recent developments in solar water-splitting photocatalysis.* Mrs Bulletin, 2011. **36**(1): p. 17-22.
12. Turner, J.A., *Sustainable hydrogen production.* Science, 2004. **305**(5686): p. 972-974.
13. Lewis, N.S. and D.G. Nocera, *Powering the planet: Chemical challenges in solar energy utilization.* Proceedings of the National Academy of Sciences of the United States of America, 2006. **103**(43): p. 15729-15735.
14. Styring, S., *Artificial photosynthesis for solar fuels.* Faraday Discussions, 2012. **155**: p. 357-376.
15. Alstrum-Acevedo, J.H., M.K. Brennaman, and T.J. Meyer, *Chemical approaches to artificial photosynthesis. 2.* Inorganic Chemistry, 2005. **44**(20): p. 6802-6827.
16. Concepcion, J.J., et al., *Chemical approaches to artificial photosynthesis.* Proceedings of the National Academy of Sciences of the United States of America, 2012. **109**(39): p. 15560-15564.
17. *University of Washington.* [cited 2013 18th February]; Available from: https://depts.washington.edu/spirolab/researchpro.html.
18. Woodhouse, M. and B.A. Parkinson, *Combinatorial approaches for the identification and optimization of oxide semiconductors for efficient solar photoelectrolysis.* Chemical Society Reviews, 2009. **38**(1): p. 197-210.
19. Grimes, C.A., O.K. Varghese, and S. Ranjan, *Light, Water, Hydrogen: The Solar Generation of Hydrogen by Water Photoelectrolysis* 2008, New York: Springer.
20. Fujishima, A. and K. Honda, *Electrochemical photolysis of water at a semiconductor electrode.* Nature, 1972. **238**(5358): p. 37-38.
21. Butler, M.A., *Photoelectrolysis and physical properties of semiconducting electrode WO_3.* Journal of Applied Physics, 1977. **48**(5): p. 1914-1920.
22. Scaife, D.E., *Oxide semiconductors in photoelectrochemical conversion of solar-energy.* Solar Energy, 1980. **25**(1): p. 41-54.
23. Kay, A., I. Cesar, and M. Graetzel, *New benchmark for water photooxidation by nanostructured alpha-Fe_2O_3 films.* Journal of the American Chemical Society, 2006. **128**(49): p. 15714-15721.
24. Yang, B., et al., *Enhanced photoelectrochemical activity of sol-gel tungsten trioxide films through textural control.* Chemistry of Materials, 2007. **19**(23): p. 5664-5672.

25. Alexander, B.D., et al., *Metal oxide photoanodes for solar hydrogen production.* Journal of Materials Chemistry, 2008. **18**(20): p. 2298-2303.
26. van de Krol, R., Y. Liang, and J. Schoonman, *Solar hydrogen production with nanostructured metal oxides.* Journal of Materials Chemistry, 2008. **18**(20): p. 2311-2320.
27. Kung, H.H., et al., *Semiconducting oxide anodes in photoassisted electrolysis of water.* Journal of Applied Physics, 1977. **48**(6): p. 2463-2469.
28. Sartoretti, C.J., et al., *Photoelectrochemical oxidation of water at transparent ferric oxide film electrodes.* Journal of Physical Chemistry B, 2005. **109**(28): p. 13685-13692.
29. Rajeshwar, K., *Hydrogen generation at irradiated oxide semiconductor-solution interfaces.* Journal of Applied Electrochemistry, 2007. **37**(7): p. 765-787.
30. Cesar, I., et al., *Influence of feature size, film thickness, and silicon doping on the performance of nanostructured hematite photoanodes for solar water splitting.* Journal of Physical Chemistry C, 2009. **113**(2): p. 772-782.
31. Solarska, R., A. Krolikowska, and J. Augustynski, *Silver Nanoparticle Induced Photocurrent Enhancement at WO3 Photoanodes.* Angewandte Chemie-International Edition, 2010. **49**(43): p. 7980-7983.
32. Sivula, K., F. Le Formal, and M. Graetzel, *Solar water splitting: progress using hematite (alpha-Fe_2O_3) photoelectrodes.* Chemsuschem, 2011. **4**(4): p. 432-449.
33. Chen, Z., et al., *Accelerating materials development for photoelectrochemical hydrogen production: Standards for methods, definitions, and reporting protocols.* Journal of Materials Research, 2010. **25**(1): p. 3-16.
34. Vidyarthi, V.S., et al., *Enhanced photoelectrochemical properties of WO_3 thin films fabricated by reactive magnetron sputtering.* International Journal of Hydrogen Energy, 2011. **36**(8): p. 4724-4731.
35. *Future Architectural Glass LLC.* [cited 2013 18th February]; Available from: http://faglass.com/?page_id=183.
36. Yang, B., et al., *Strong photoresponse of nanostructured tungsten trioxide films prepared via a sol-gel route.* Journal of Materials Chemistry, 2007. **17**(26): p. 2722-2729.
37. Cristino, V., et al., *Efficient photoelectrochemical water splitting by anodically grown WO_3 electrodes.* Langmuir, 2011. **27**(11): p. 7276-7284.
38. Stepanovich, A., et al., *Combinatorial development of nanoporous WO_3 thin film photoelectrodes for solar water splitting by dealloying of binary alloys.* International Journal of Hydrogen Energy, 2012. **37**(16): p. 11618-11624.
39. Wang, F.G., C. Di Valentin, and G. Pacchioni, *Doping of WO_3 for photocatalytic water splitting: hints from density functional theory.* Journal of Physical Chemistry C, 2012. **116**(16): p. 8901-8909.
40. Hwang, D.W., et al., *Mg-doped WO_3 as a novel photocatalyst for visible light-induced water splitting.* Catalysis Letters, 2002. **80**(1-2): p. 53-57.
41. Behnajady, M.A., B. Alizade, and N. Modirshahla, *Synthesis of Mg-doped TiO_2 nanoparticles under different conditions and its photocatalytic activity.* Photochemistry and Photobiology, 2011. **87**(6): p. 1308-1314.
42. Nah, Y.C., et al., *Nitrogen doping of nanoporous WO_3 layers by NH_3 treatment for increased visible light photoresponse.* Nanotechnology, 2010. **21**(10).
43. Liu, Z., et al., *Dealloying derived synthesis of W nanopetal films and their transformation into WO_3.* Journal of Physical Chemistry C, 2008. **112**(5): p. 1391-1395.
44. IUPAC. *Compendium of Chemical Terminology*, 2nd ed. (the "Gold Book"). Compiled by A. D. McNaught and A. Wilkinson. Blackwell Scientific Publications, Oxford (1997). XML on-line corrected version: http://goldbook.iupac.org (2006-) created by M. Nic, J. Jirat, B. Kosata; updates compiled by A. Jenkins. ISBN 0-9678550-9-8. doi:10.1351/goldbook. Last update: 2012-08-19; version: 2.3.2.
45. Guy, A.G. and R. Ciach, *Wprowadzenie do nauki o materiałach* 1977: Państwowe Wydaw. Naukowe.
46. West, J.M., *Basic Corrosion and Oxidation* 1992: Ellis Horwood.
47. Baszkiewicz, J. and M. Kamiński, *Podstawy korozji materiałów* 1997: Oficyna Wydawnicza Politechniki Warszawskiej.

48. Bates, R.G. and J.B. MacAskill, *Standard potential of the silver-silver chloride electrode.* Pure & Applied Chem, 1979. **50**: p. 1701-1706.
49. Bates, R.G., *Treatise on Analytical Chemistry*, I.M. Kolthoff and P.J. Elving, Editors. 1978, Wiley: New York. p. 793.
50. Pourbaix, M., *Atlas of Electrochemical Equilibria in Aqueous Solutions* 1966, New York: Pergamon Press.
51. Fekry, A.M. and R.H. Tammam, *Corrosion and impedance studies on magnesium alloy in oxalate solution.* Materials Science and Engineering B-Advanced Functional Solid-State Materials, 2011. **176**(10): p. 792-798.
52. Pugh, D.V., A. Dursun, and S.G. Corcoran, *Formation of nanoporous platinum by selective dissolution of Cu from $Cu_{0.75}Pt_{0.25}$.* Journal of Materials Research, 2003. **18**(1): p. 216-221.
53. Zhu, S.L., et al., *Ti oxide nano-porous surface structure prepared by dealloying of Ti-Cu amorphous alloy.* Electrochemistry Communications, 2011. **13**(3): p. 250-253.
54. Gerischer, H. and H. Rickert, *Uber das elektrochemische Verhalten von Kupfer-Gold-Legierungen und den Mechanismus der Spannungskorrosion.* Zeitschrift Fur Metallkunde, 1955. **46**(9): p. 681-689.
55. Tischer, R.P. and H. Gerischer, *Elektrolytische Auflosung von Gold-Silber-Legirungen und die Frage der Resistenzgrenzen.* Zeitschrift Fur Elektrochemie, 1958. **62**(1): p. 50-60.
56. Gerischer, H., *Korrosionsshutz durch Lagieren*, in *Verlag Chemie*, T. Heumann, Editor 1962, Verlag Chemie: Weinheim.
57. Marshakov, I.K., *Selective corrosion of alloys.* Soros Educational Journal, 2000. **6**(4): p. 57-62.
58. Forty, A.J., *Micromorphological studies of the corrosion of gold alloys.* Gold Bulletin, 1981. **14**: p. 25-35.
59. Sieradzki, K., *Curvature effects in alloy dissolution.* Journal of the Electrochemical Society, 1993. **140**(10): p. 2868-2872.
60. Sieradzki, K., et al., *The relationship between dealloying and transgranular stress-corrosion cracking of Cu-Zn and Cu-Al alloys.* Journal of the Electrochemical Society, 1987. **134**(7): p. 1635-1639.
61. Lu, X., et al., *Dealloying of Au-Ag thin films with a composition gradient: Influence on morphology of nanoporous Au.* Thin Solid Films, 2007. **515**(18): p. 7122-7126.
62. A. Wittstock, J. Biener, and M. Bäumer, *Nanoporous Gold: A Novel Catalyst with Tunable Properties*, in *Dealloying Process and Related Synthetic Opportunities*, T. P. Moffat, J. Erlebacher, and R.C. Newman, Editors. 2010, The Electrochemical Society: Pennington, New Jersey, USA.
63. L. Chen, X. Lang, and M. Chen, *Dealloyed Nanoporous Metals*, in *Nanoporous Materials: Synthesis and Applications*, Q. Xu, Editor 2013, Taylor & Francis Group, LLC: Boca Raton, FL, USA. p. 125-182.
64. Pickering, H.W. and P.J. Byrne, *On Preferential Anodic Dissolution of Alloys in the Low-Current Region and the Nature of the Critical Potential.* J. Electrochemical Soc., 1971. **118**(2): p. 209-215.
65. Pryor, M.J. and K.-K. Giam, *The Effect of Arsenic on the Dealloying of α-Brass.* J. Electrochem. Soc., 1982. **129**(10): p. 2157-2163.
66. Stratmann, M. and M. Rohwerder, *Materials science - A pore view of corrosion.* Nature, 2001. **410**(6827): p. 420-423.
67. Renner, F.U., *In-situ X-ray study of the initial electrochemical corrosion of Cu-Au(111)*, 2004, University Stuttgart.
68. Boettcher, A., G. Haase, and R. Thun, *Strukturuntersuchung von Mehrstoffsystemen durch kinematische Elektronenbeugung.* Zeitschrift Fur Metallkunde, 1955. **46**(5): p. 386-400.
69. Kennedy, K., et al., *Rapid method for determining ternary-alloy phase diagrams.* Journal of Applied Physics, 1965. **36**(12): p. 3808-3810.
70. Hanak, J.J., *Multiple-sample-concept in materials research - synthesis, compositional analysis and testing of entire multicomponent systems.* Journal of Materials Science, 1970. **5**(11): p. 964-&.
71. Hanak, J.J., R.K. Wehner, and H.W. Lehmann, *Calculation of deposition profiles and compositional analysis of cosputtered films.* Journal of Applied Physics, 1972. **43**(4): p. 1666-&.
72. Xiang, X.D., et al., *A combinatorial approach to materials discovery.* Science, 1995. **268**(5218): p. 1738-1740.

73. Briceno, G., et al., *A class of cobalt oxide magnetoresistance materials discovered with combinatorial synthesis.* Science, 1995. **270**(5234): p. 273-275.
74. Hyett, G., M.A. Green, and I.P. Parkin, *The use of combinatorial chemical vapor deposition in the synthesis of $Ti_{(3-delta)}O_4N$ with $0.06 < delta < 0.25$: A titanium oxynitride phase Isostructural to anosovite.* Journal of the American Chemical Society, 2007. **129**(50): p. 15541-15548.
75. Hyett, G., M. Green, and I.P. Parkin, *X-ray diffraction area mapping of preferred orientation and phase change in TiO_2 thin films deposited by chemical vapor deposition.* Journal of the American Chemical Society, 2006. **128**(37): p. 12147-12155.
76. Kafizas, A., C. Crick, and I.P. Parkin, *The combinatorial atmospheric pressure chemical vapour deposition (cAPCVD) of a gradating substitutional/interstitial N-doped anatase TiO_2 thin-film; UVA and visible light photocatalytic activities.* Journal of Photochemistry and Photobiology a-Chemistry, 2010. **216**(2-3): p. 156-166.
77. Kafizas, A., et al., *Simple method for the rapid simultaneous screening of photocatalytic activity over multiple positions of self-cleaning films.* Physical Chemistry Chemical Physics, 2009. **11**(37): p. 8367-8375.
78. Kafizas, A., C.W. Dunnill, and I.P. Parkin, *Combinatorial atmospheric pressure chemical vapour deposition (cAPCVD) of niobium doped anatase; effect of niobium on the conductivity and photocatalytic activity.* Journal of Materials Chemistry, 2010. **20**(38): p. 8336-8349.
79. Wang, Q., F.Z. Liu, and D.X. Han, *High-throughput chemical vapor deposition system and thin-film silicon library.* Macromolecular Rapid Communications, 2004. **25**(1): p. 326-329.
80. Hansel, H., et al., *Combinatorial study of the long-term stability of organic thin-film solar cells.* Applied Physics Letters, 2002. **81**(11): p. 2106-2108.
81. Wang, Q., *Combinatorial hot-wire CVD approach to exploring thin-film Si materials and devices.* Thin Solid Films, 2003. **430**(1-2): p. 78-82.
82. Dam, B., et al., *Combinatorial thin film methods for the search of new lightweight metal hydrides.* Scripta Materialia, 2007. **56**(10): p. 853-858.
83. Gremaud, R., et al., *Hydrogenography: An optical combinatorial method to find new light-weight hydrogen-storage materials.* Advanced Materials, 2007. **19**(19): p. 2813-+.
84. Olk, C.H., et al., *Combinatorial preparation and infrared screening of hydrogen sorbing metal alloys.* Journal of Applied Physics, 2003. **94**(1): p. 720-725.
85. Ludwig, A., et al., *Opto-mechanical characterization of hydrogen storage properties of Mg-Ni thin film composition spreads.* Applied Surface Science, 2007. **254**(3): p. 682-686.
86. Garsuch, A., et al., *The effect of boron doping into Co-C-N and Fe-C-N electrocatalysts on the oxygen reduction reaction.* Electrochimica Acta, 2008. **53**(5): p. 2423-2429.
87. Cooper, J.S. and P.J. McGinn, *Combinatorial screening of fuel cell cathode catalyst compositions.* Applied Surface Science, 2007. **254**(3): p. 662-668.
88. Easton, E.B., et al., *Magnetron sputtered Fe-C-N, Fe-C, and C-N based oxygen reduction electrocatalysts.* Journal of the Electrochemical Society, 2008. **155**(6): p. B547-B557.
89. Whitacre, J.F., T.I. Valdez, and S.R. Narayanan, *A high-throughput study of PtNiZr catalysts for application in PEM fuel cells.* Electrochimica Acta, 2008. **53**(10): p. 3680-3689.
90. Cooper, J.S. and P.J. McGinn, *Combinatorial screening of thin film electrocatalysts for a direct methanol fuel cell anode.* Journal of Power Sources, 2006. **163**(1): p. 330-338.
91. Lu, G.J., J.S. Cooper, and P.J. McGinn, *SECM characterization of Pt-Ru-WC and Pt-Ru-Co ternary thin film combinatorial libraries as anode electrocatalysts for PEMFC.* Journal of Power Sources, 2006. **161**(1): p. 106-114.
92. Todd, A.D.W., R.E. Mar, and J.R. Dahn, *Combinatorial study of tin-transition metal alloys as negative electrodes for lithium-ion batteries.* Journal of the Electrochemical Society, 2006. **153**(10): p. A1998-A2005.

93. Dahn, J.R., R.E. Mar, and A. Abouzeid, *Combinatorial study of $Sn_{1-x}Co_x$ (0 < x < 0.6) and $Sn_{0.55}Co_{0.45}$ (1-y)C_y (0 < y < 0.5) alloy negative electrode materials for Li-ion batteries*. Journal of the Electrochemical Society, 2006. **153**(2): p. A361-A365.
94. Borgia, C., et al., *A combinatorial study on the influence of Cu addition, film thickness and heat treatment on phase composition, texture and mechanical properties of Ti-Ni shape memory alloy thin films*. Thin Solid Films, 2010. **518**(8): p. 1897-1913.
95. Cui, J., et al., *Combinatorial search of thermoelastic shape-memory alloys with extremely small hysteresis width*. Nature Materials, 2006. **5**(4): p. 286-290.
96. Zarnetta, R., et al., *Combinatorial study of phase transformation characteristics of a Ti-Ni-Pd shape memory thin film composition spread in view of microactuator applications*. Applied Surface Science, 2007. **254**(3): p. 743-748.
97. Lobel, R., et al., Materials Science and Engineering: A, 2008. **151**: p. 481-482.
98. Buenconsejo, P.J.S., R. Zarnetta, and A. Ludwig, *The effects of grain size on the phase transformation properties of annealed (Ti/Ni/W) shape memory alloy multilayers*. Scripta Materialia, 2011. **64**(11): p. 1047-1050.
99. Buenconsejo, P.J.S., et al., *A new prototype two-phase (TiNi)-(beta-W) SMA system with tailorable thermal hysteresis*. Advanced Functional Materials, 2011. **21**(1): p. 113-118.
100. Hamann, S., et al., *The ferromagnetic shape memory system Fe-Pd-Cu*. Acta Materialia, 2010. **58**(18): p. 5949-5961.
101. Wakisaka, T., et al., *$Sr_2Rh_{1-x}Ru_xO_4$ (0 <= x <= 1) composition-spread film growth on a temperature-gradient substrate by pulsed laser deposition*. Applied Surface Science, 2004. **223**(1-3): p. 264-267.
102. Wang, J.S., et al., *Identification of a blue photoluminescent composite material from a combinatorial library*. Science, 1998. **279**(5357): p. 1712-1714.
103. Kukuruznyak, D.A., et al., *Electrical screening of ternary $NiO-Mn_2O_3-Co_3O_4$ composition spreads*. Applied Surface Science, 2006. **252**(10): p. 3828-3832.
104. Martel, A., et al., *Physical properties of transparent conducting Cd-Te-In-O thin films Outlining a thermodynamic system for transparent conducting oxides*. Thin Solid Films, 2009. **518**(1): p. 413-418.
105. Joshi, U.S. and H. Koinuma, *Binary composition spread approach for parallel pulsed laser deposition synthesis and highthroughput characterization of transparent and semiconducting oxide thin films*. Indian Journal of Pure & Applied Physics, 2007. **45**(1): p. 62-65.
106. Perkins, J.D., et al., *Combinatorial studies of Zn-Al-O and Zn-Sn-O transparent conducting oxide thin films*. Thin Solid Films, 2002. **411**(1): p. 152-160.
107. Guan, Y.F., et al., *Combinatorial synthesis and characterization of magnetic $Fe_{(x)}Al_{(1-x)}N_{(y)}O_{(1-y)}$ thin films*. Thin Solid Films, 2008. **516**(18): p. 6063-6070.
108. Joshi, U.S., et al., *Combinatorial fabrication and magnetic properties of homoepitaxial Co and Li co-doped NiO thin-film nanostructures*. Journal of Magnetism and Magnetic Materials, 2009. **321**(21): p. 3595-3599.
109. Matsumoto, Y., et al., *Structural control and combinatorial doping of titanium dioxide thin films by laser molecular beam epitaxy*. Applied Surface Science, 2002. **189**(3-4): p. 344-348.
110. Matsumoto, Y., et al., *Room-temperature ferromagnetism in transparent transition metal-doped titanium dioxide*. Science, 2001. **291**(5505): p. 854-856.
111. He, J., et al., *Room temperature ferromagnetic n-type semiconductor in $(In_{1-x}Fe_x)_2O_3$-sigma*. Applied Physics Letters, 2005. **86**(5).
112. Buschbeck, J., et al., *Correlation of phase transformations and magnetic properties in annealed epitaxial Fe-Pd magnetic shape memory alloy films*. Journal of Applied Physics, 2010. **107**(11).
113. Brunken, H., et al., *Microstructure and magnetic properties of FeCo/Ti thin film multilayers annealed in nitrogen*. Thin Solid Films, 2010. **519**(2): p. 770-774.
114. Zhong, L.J., et al., *Combinatorial CVD of ZrO_2 or HfO_2 compositional spreads with SiO_2 for high kappa dielectrics*. Journal of Materials Chemistry, 2004. **14**(21): p. 3203-3209.

115. Guozhen, L., et al., *Microstructural and dielectric properties of $Ba_{0.6}Sr_{0.4}Ti_{1-x}Zr_xO_3$ based combinatorial thin film capacitors library.* Journal of Applied Physics, 2010. **108**(11): p. 114108 (6 pp.)-114108 (6 pp.)114108 (6 pp.).
116. van Dover, R.B., et al., *A high-throughput search for electronic materials thin-film dielectrics.* Biotechnology and Bioengineering, 1999. **61**(4): p. 217-225.
117. Xia, B., et al., *Combinatorial CVD of zirconium, hafnium, and tin dioxide mixtures for applications as high-k materials.* Chemical Vapor Deposition, 2004. **10**(4): p. 195-200.
118. Fowlkes, J.D., J.M. Fitz-Gerald, and P.D. Rack, *Ultraviolet emitting $(Y_{1-x}Gd_x)_2O_3$-delta thin films deposited by radio frequency magnetron sputtering: Combinatorial modeling, synthesis, and rapid characterization.* Thin Solid Films, 2006. **510**(1-2): p. 68-76.
119. Muramatsu, Y., et al., *Combinatorial synthesis and screening for blue luminescent pi-conjugated polymer thin film.* Applied Surface Science, 2002. **189**(3-4): p. 319-326.
120. Deng, Y., Y.F. Guan, and P.D. Rack, *Combinatorial synthesis and sputter parameter optimization of chromium-doped yttrium aluminum garnet photoluminescent thin films.* Thin Solid Films, 2006. **515**(4): p. 1721-1726.
121. Danielson, E., et al., *A rare-earth phosphor containing one-dimensional chains identified through combinatorial methods.* Science, 1998. **279**(5352): p. 837-839.
122. Sano, H., et al., *A combinatorial approach to the discovery and optimization of $YCa_4O(BO_3)_{(3)}$-based luminescent materials.* Applied Surface Science, 2006. **252**(7): p. 2493-2496.
123. Sano, H., et al., *$ReCa_4O(BO3)_{(3)}$ thin films as new Luminescent materials screened by the combinatorial method.* Journal of Physics and Chemistry of Solids, 2005. **66**(11): p. 2112-2115.
124. Miyata, T., Y. Mochizuki, and T. Minami, *High-luminance EL devices using $Zn_2Si_{1-x}Ge_xO_4$: Mn thin films prepared by combinatorial deposition by RF magnetron sputtering with subdivided powder targets.* Ieice Transactions on Electronics, 2005. **E88C**(11): p. 2065-2069.
125. Peak, J.D., C.L. Melcher, and P.D. Rack, *Combinatorial thin film synthesis of cerium doped scintillation materials in the lutetium oxide-silicon oxide system.* Ieee Transactions on Nuclear Science, 2008. **55**(3): p. 1480-1483.
126. Simon, C.G., et al., *Combinatorial screening of cell proliferation on poly(D,L-lactic acid)/poly(D,L-lactic acid) blends.* Biomaterials, 2005. **26**(34): p. 6906-6915.
127. Hoogenboom, R., M.A.R. Meier, and U.S. Schubert, *Combinatorial methods, automated synthesis and high-throughput screening in polymer research: Past and present.* Macromolecular Rapid Communications, 2003. **24**(1): p. 16-32.
128. Walter, H., et al., *Combinatorial Approach for Fast Screening of Functional Materials.* Journal of Polymer Science Part B-Polymer Physics, 2010. **48**(14): p. 1587-1593.
129. Ludwig, A., et al., *Development of multifunctional thin films using high-throughput experimentation methods.* International Journal of Materials Research, 2008. **99**(10): p. 1144-1149.
130. Hassel, A.W. and M.M. Lohrengel, *The scanning droplet cell and its application to structured nanometer oxide films on aluminium.* Electrochimica Acta, 1997. **42**(20-22): p. 3327-3333.
131. Takeuchi, I., et al., *Identification of novel compositions of ferromagnetic shape-memory alloys using composition spreads.* Nature Materials, 2003. **2**(3): p. 180-184.
132. Wu, K.H., et al., *Improvement of spatial resolution for local Seebeck coefficient measurements by deconvolution algorithm.* Review of Scientific Instruments, 2009. **80**(10).
133. Ludwig, A., et al., *MEMS tools for combinatorial materials processing and high-throughput characterization.* Measurement Science & Technology, 2005. **16**(1): p. 111-118.
134. Smith, D.O., *Static and dynamic behavior of thin permalloy films.* Journal of Applied Physics, 1958. **29**(3): p. 264-273.
135. Knorr, T.G. and R.W. Hoffman, *Dependence of geometric magnetic anisotropy in thin iron films.* Physical Review, 1959. **113**(4): p. 1039-1046.
136. Liu, F., et al., *The growth of nanoscale structured iron films by glancing angle deposition.* Journal of Applied Physics, 1999. **85**(8): p. 5486-5488.

137. Varghese, O.K. and C.A. Grimes, *Appropriate strategies for determining the photoconversion efficiency of water photo electrolysis cells: A review with examples using titania nanotube array photoanodes.* Solar Energy Materials and Solar Cells, 2008. **92**(4): p. 374-384.
138. Reiter, S., et al., *Redox modification of proteins using sequential-parallel electrochemistry in microtiter plates.* Analyst, 2001. **126**(11): p. 1912-1918.
139. Reiter, S., et al., *An electrochemical robotic system for the optimization of amperometric glucose biosensors based on a library of cathodic electrodeposition paints.* Macromolecular Rapid Communications, 2004. **25**(1): p. 348-354.
140. Ryabova, V., et al., *Robotic sequential analysis of a library of metalloporphyrins as electrocatalysts for voltammetric nitric oxide sensors.* Analyst, 2005. **130**(9): p. 1245-1252.
141. Eckhard, K., et al., *Spatially resolved mass spectrometry as a fast semi-quantitative tool for testing heterogeneous catalyst libraries under reducing stagnant-point flow conditions.* Applied Catalysis a-General, 2005. **281**(1-2): p. 115-120.
142. Erichsen, T., et al., *Combinatorial microelectrochemistry: Development and evaluation of an electrochemical robotic system.* Review of Scientific Instruments, 2005. **76**(6).
143. Lohrengel, M.M., A. Moehring, and M. Pilaski, *Electrochemical surface analysis with the scanning droplet cell.* Fresenius Journal of Analytical Chemistry, 2000. **367**(4): p. 334-339.
144. Knotek, O., F. Loffler, and A. Barimani, *Interface Stabilization of W-N Coatings by Chromium Alloying.* Interface Dynamics and Growth, ed. K.S. Liang, et al. Vol. 237. 1992, Pittsburgh: Materials Research Soc. 673-678.
145. Cavaleiro, A., B. Trindade, and M.T. Vieira, *Chapter 7: The Influence of the Addition of a Third Element on the Structure and Mechanical Properties of Transition-Metal-Based Nanostructured Hard Films: Part I—Nitrides*, in *Nanostructured Coatings*, A. Cavaleiro and J.T.M.D. Hosson, Editors. 2006, Springer Science+Business Media, LLC: New York.
146. Inoue, A. and A. Takeuchi, *Recent progress in bulk glassy alloys.* Materials Transactions, 2002. **43**(8): p. 1892-1906.
147. Arun Pratap and A.T. Patel, *Crystallization Kinetics of Metallic Glasses*, in *Advances in Crystallization Processes*, D.Y. Mastai, Editor 2012, InTech: Shanghai. p. 107-126.
148. Erlebacher, J., et al., *Evolution of nanoporosity in dealloying.* Nature, 2001. **410**(6827): p. 450-453.
149. Erlebacher, J. and K. Sieradzki, *Pattern formation during dealloying.* Scripta Materialia, 2003. **49**(10): p. 991-996.
150. Erlebacher, J., *An atomistic description of dealloying - Porosity evolution, the critical potential, and rate-limiting behavior.* Journal of the Electrochemical Society, 2004. **151**(10): p. C614-C626.
151. Mullins, W.W., *Flattening of a nearly plane solid surface due to capillarity* Journal of Applied Physics, 1959. **30**(1): p. 77-83.
152. Gillet, M., et al., *The structure and electrical conductivity of vacuum-annealed WO3 thin films.* Thin Solid Films, 2004. **467**(1-2): p. 239-246.
153. Aburada, T., J.M. Fitz-Gerald, and J.R. Scully, *Pitting and dealloying of solute-rich Al-Cu-Mg-based amorphous alloys: effect of alloying with minor concentrations of nickel.* Journal of the Electrochemical Society, 2011. **158**(9): p. C253-C265.
154. Robbie, K. and M.J. Brett, *Sculptured thin films and glancing angle deposition: Growth mechanics and applications.* Journal of Vacuum Science & Technology a-Vacuum Surfaces and Films, 1997. **15**(3): p. 1460-1465.
155. Haghirigosnet, A.M., et al., *Stress and microstructure in tungsten sputterend thin films.* Journal of Vacuum Science & Technology a-Vacuum Surfaces and Films, 1989. **7**(4): p. 2663-2669.
156. Okeefe, M.J. and J.T. Grant, *Phase transformation of sputter deposited tungsten thin films with A-15 structure.* Journal of Applied Physics, 1996. **79**(12): p. 9134-9141.
157. Weerasekera, I.A., et al., *Structure and stability of sputter deposited beta-tungsten thin films.* Applied Physics Letters, 1994. **64**(24): p. 3231-3233.

158. Shen, Y.G. and Y.W. Mai, *Influences of oxygen on the formation and stability of A15 beta-W thin films.* Materials Science and Engineering a-Structural Materials Properties Microstructure and Processing, 2000. **284**(1-2): p. 176-183.
159. Baumann, F.H., et al., *Multiscale modeling of thin-film deposition: Applications to Si device processing.* Mrs Bulletin, 2001. **26**(3): p. 182-189.
160. Karabacak, T., et al., *beta-phase tungsten nanorod formation by oblique-angle sputter deposition.* Applied Physics Letters, 2003. **83**(15): p. 3096-3098.
161. Karabacak, T., et al., *Phase transformation of single crystal beta-tungsten nanorods at elevated temperatures.* Thin Solid Films, 2005. **493**(1-2): p. 293-296.

List of abbreviations and symbols

ρ – electrical resistivity;

θ – the angle between the incident X-ray and the scattering atomic planes;

λ – wavelength;

a – chemical activity for the relevant species;

ADC – analog to digital converter;

AM1 – air mass coefficient;

at. % – atomic percent;

c – concentration;

ca. – circa (from Latin, meaning "around, about");

d – inter-planar lattice distance;

DC – direct current;

e^- – electron;

e.g. – exempli gratia (from Latin, meaning "for example");

E^0 – standard half-cell reduction potential;

EDX – energy dispersive X-ray analysis;

F – Faraday constant, the number of coulombs per mole of electrons: $F = 9.648\,533\,99 \times 10^4$ C mol^{-1};

FIB – focused ion beam;

GLAD – glancing angle deposition;

h – film thickness;

M – metal;

n – an integer number;

N_i – mole fraction of the component "i";

OSDC – optical scanning droplet cell;

p – pressure;

pDC – pulsed direct current;

PEC – photoelectrochemical;

pH – a measure of the acidity/basicity of an aqueous solution;

PTFE – polytetrafluoroethylene;

PVD – physical vapor deposition;

R – universal gas constant: $R = 8.314\,472$ $J\,K^{-1}\,mol^{-1}$;

RF – radio frequency;

SAED – selective area electron diffraction;

SEM – scanning electron microscopy;

SHE – standard hydrogen electrode;

T – absolute temperature;

t_d – time of deposition;

TEM – transmission electron microscope;

UV – ultraviolet;

vs. – versus;

wt. % – weight percent;

XRD – X-ray diffraction;

z – the number of moles of electrons transferred in the half-reaction;

γ_X – activity coefficient of relevant species.

Journal publications contributing to the thesis:

1. **A.Stepanovich**, K.Sliozberg, W.Schuhmann, A.Ludwig "Combinatorial development of nanoporous WO_3 thin film photoelectrodes for solar water splitting by dealloying of binary alloys" International Journal of Hydrogen Energy, 2012, 37, 11618-11624.

2. C.Khare, **A.Stepanovich**, P.J.Buenconsejo, A.Ludwig "Synthesis of WO_3 nanoblades by dealloying glancing angle deposited W-Fe nanocolumnar thin films"Nanotechnology, 2014, 25(20), 205606.

Conferences and meetings:

1. **A.Stepanovich**, K.Sliozberg, W.Schuhmann, A.Ludwig "Combinatorial development and investigation of nanoporous WO_3 films for solar water splitting by dealloying of binary alloys with further oxidation" 8th International Conference "Porous Semiconductors – Science and Technology", Malaga, Spain, March 25-30, 2012. (oral presentation)

2. **A.Stepanovich**, K.Sliozberg, W.Schuhmann, A.Ludwig "Combinatorial development of nanoporous semiconductor thin film photoelectrodes for solar water splitting by dealloying of binary and ternary alloys", 10th Materials Day, Bochum, Germany, November 8-9, 2012. (poster presentation)

3. **A.Stepanovich**, K.Sliozberg, W.Schuhmann, A.Ludwig "Combinatorial development of porous semiconductor thin film electrodes for solar water splitting by dealloying of binary and ternary alloys", 2nd International Conference on Materials for Energy "EnMat II", Karlsruhe, Germany, May 12-16, 2013. (oral presentation)

4. C.Khare, **A.Stepanovich**, K.Sliozberg, P.J.Buenconsejo, W.Schuhmann, A.Ludwig "Combinatorial development of Fe-doped WO_3 nanostructures for photoelectrochemical solar water-splitting", European Congress and Exhibition on Advanced Materials and Processes "EUROMAT 2013", Sevilla, Spain, September 8-13, 2013. (oral presentation)

I want morebooks!

Buy your books fast and straightforward online - at one of the world's fastest growing online book stores! Environmentally sound due to Print-on-Demand technologies.

Buy your books online at
www.get-morebooks.com

Kaufen Sie Ihre Bücher schnell und unkompliziert online – auf einer der am schnellsten wachsenden Buchhandelsplattformen weltweit! Dank Print-On-Demand umwelt- und ressourcenschonend produziert.

Bücher schneller online kaufen
www.morebooks.de

OmniScriptum Marketing DEU GmbH
Heinrich-Böcking-Str. 6-8
D - 66121 Saarbrücken

Telefax: +49 681 93 81 567-9

info@omniscriptum.de
www.omniscriptum.de

Printed by Books on Demand GmbH, Norderstedt / Germany